AF357333

Extra Virgin Olive Oil Quality, Safety, and Authenticity

Extra Virgin Olive Oil Quality, Safety, and Authenticity

Editor

Theodoros Varzakas

MDPI • Basel • Beijing • Wuhan • Barcelona • Belgrade • Manchester • Tokyo • Cluj • Tianjin

Editor
Theodoros Varzakas
Food Science and Technology
University of the Peloponnese
Kalamata
Greece

Editorial Office
MDPI
St. Alban-Anlage 66
4052 Basel, Switzerland

This is a reprint of articles from the Special Issue published online in the open access journal *Foods* (ISSN 2304-8158) (available at: www.mdpi.com/journal/foods/special_issues/virgin_olive_oil_quality).

For citation purposes, cite each article independently as indicated on the article page online and as indicated below:

LastName, A.A.; LastName, B.B.; LastName, C.C. Article Title. *Journal Name* **Year**, *Volume Number*, Page Range.

ISBN 978-3-0365-1844-2 (Hbk)
ISBN 978-3-0365-1843-5 (PDF)

Contents

About the Editor

Theodoros Varzakas

Theodoros Varzakas is a Senior Full Professor at the Department of Food Science and Technology, University of Peloponnese, Greece, specializing in issues of food technology, food processing/engineering, and food quality and safety; Section Editor in Chief in the journal *Foods in Food Security and Sustainability* (2020-); Ex Editor in Chief of *Current Research in Nutrition and Food Science* (2015-2019); and Reviewer and member of the editorial board in many international journals. He has written more than 200 research papers and reviews and has presented more than 160 papers and posters at national and international conferences. He has written and edited six books in Greek and six in English on sweeteners, biosensors, food engineering, and food processing published by CRC. He has participated in many European and national research programs as coordinator or scientific member.

Preface to "Extra Virgin Olive Oil Quality, Safety, and Authenticity"

The prevention and bioactivity effects associated with the so-called "Mediterranean diet"make olive oil the most consumed edible fat in the food intake of the Mediterranean basin.

The road to quality demands that legislation should be followed. Hence, official European Union classifications such as protected designation of origin (PDO), protected geographical indication (PGI), etc. guarantee the quality and the origin of the labeled foodstuff.

The profiling of volatile components and the aroma of olive oil are key factors in the quality dimension and are affected by various factors and conditions such as cultivar; atmospheric, pedologic, and fostering conditions; the ripening degree; olive and oil storage; and the technology of oil extraction from drupes, as well as the quality of the pre-extraction procedures.

In extra virgin olive oil production, as in all kinds of production, the maintenance of high quality standards is assured by the olive fruits'and the final products'quality. Modern milling technologies can aid in the direction of quality and safety and thus can be employed in the production of extra virgin olive oil (EVOO), which can be directly consumed without any further manipulation. The overall quality of EVOO should be determined by quality characteristics including sensory analysis, stability, and nutritional value and safety (microbiology, absence of contaminants and toxins), along with authenticity.

Food authenticity issues are very important for the food industry due to legislation aspects, economics, quality specifications and conformance, safety concerns, and religious matters. Authentic EVOO should comply with the producer's declaration regarding the quality of olive fruits, natural components, the absence of extraneous substances, production technology, the geographical and botanical origin, the production year, and the genetic identity. Hence, olive oil authenticity can be implemented by the validation of the application of accurate specifications for olive fruits and the selection of trustworthy suppliers with a quality assurance system in place. Authenticity methodologies will avoid adulteration but will also aid the control of accidental contaminations, e.g., in factories, where several oils are produced or used at the same time, or cross-contaminations.

Theodoros Varzakas
Editor

Editorial

Extra Virgin Olive Oil (EVOO): Quality, Safety, Authenticity, and Adulteration

Theodoros Varzakas 

Department Food Science and Technology, University of the Peloponnese, 24100 Kalamata, Greece;
t.varzakas@uop.gr

Citation: Varzakas, T. Extra Virgin Olive Oil (EVOO): Quality, Safety, Authenticity, and Adulteration. *Foods* **2021**, *10*, 995. https://doi.org/10.3390/foods10050995

Received: 25 April 2021
Accepted: 30 April 2021
Published: 2 May 2021

Publisher's Note: MDPI stays neutral with regard to jurisdictional claims in published maps and institutional affiliations.

The prevention and bioactivity effects associated with the so-called "Mediterranean diet" make olive oil the most consumed edible fat in the food intake of the Mediterranean basin.

The road to quality demands that legislation should be followed. Hence, official European Union classifications such as protected designation of origin (PDO), protected geographical indication (PGI), etc. guarantee the quality and the origin of the labeled foodstuff.

The profiling of volatile components and the aroma of olive oil are key factors in the quality dimension and are affected by various factors and conditions such as cultivar; atmospheric, pedologic, and fostering conditions; the ripening degree; olive and oil storage; and the technology of oil extraction from drupes, as well as the quality of the pre-extraction procedures.

In extra virgin olive oil production, as in all kinds of production, the maintenance of high quality standards is assured by the olive fruits' and the final products' quality. Modern milling technologies can aid in the direction of quality and safety, and thus can be employed in the production of extra virgin olive oil (EVOO), which can be directly consumed without any further manipulation. The overall quality of EVOO should be determined by quality characteristics including sensory analysis, stability, and nutritional value and safety (microbiology, absence of contaminants and toxins), along with authenticity.

Food authenticity issues are very important for the food industry due to legislation aspects, economics, quality specifications and conformance, safety concerns, and religious matters. Authentic EVOO should comply with the producer's declaration regarding the quality of olive fruits, natural components, the absence of extraneous substances, production technology, the geographical and botanical origin, the production year, and the genetic identity. Hence, olive oil authenticity can be implemented by the validation of the application of accurate specifications for olive fruits and the selection of trustworthy suppliers with a quality assurance system in place. Adulteration is usually carried out for economical purposes such as the increase of the bulk volume, the increase in the yield of a process, or the advancing of the quality of a product of inferior quality by mixing. Authenticity methodologies will avoid adulteration but will also aid the control of accidental contaminations, e.g., in factories, where several oils are produced or used at the same time, or cross-contaminations.

In this Special Issue, both issues of olive fruits and EVOO have been investigated.

Virgin olive oil production parameters, composition, and quality were determined after storage for seven days at room temperature (RT), refrigerated, and frozen storage prior to oil production derived from post-harvest olive fruits (Istarska bjelica (IB) and Rosinjola (RO)). It was found that lower temperatures delayed the post-harvest maturation of IB fruits. Storage at RT maintained the highest oil yield and extractability index. Storage at RT increased the content of waxes, while the lower temperatures partially suppressed this phenomenon. Refrigerated storage preserved the concentration of the most phenolic compounds. Refrigeration seems to be the most suitable option for prolonged fruit storage [1].

Another study helped to prevent EVOO from fraud and adulteration and evaluated olive oil geographical origin by an open source visible and near infra-red (VIS-NIR) spectrophotometer [2]. They analyzed 67 Italian and 25 foreign EVOO samples, and multivariate analysis of variance (MANOVA) results reported significant differences ($p < 0.001$) between the Italian and foreign EVOO VIS-NIR matrices. They also employed an artificial neural network (ANN) model with an external test.

The next paper deals with the adulteration and authentication of extra virgin olive oil (EVOO) using vibrational spectroscopy signatures combined with pattern recognition analysis [3]. Oils were characterized by quality parameters such as fatty acid profile, free fatty acids (FFA), peroxide value (PV), pyropheophytins (PPP), and total polar compounds (TPC). Both techniques identified EVOO adulteration with vegetable oils, but Raman spectroscopy showed limited resolution detecting VOO/OO tampering. Excellent correlation was shown by partial least squares regression models.

The next work (H2020 OLEUM project) [4] represents the first published attempt to verify some of the recommended quality control tools for increasing harmonization among sensory panels. A new "decision tree" scheme was developed, and some IOC quality control procedures were applied. The adoption of these tools allowed for the reliable classification of 289 out of 334 VOOs. A "formative reassessment" was necessary to control misalignments. The authors reported the need to adopt new stable and reproducible reference materials in order to improve the panel's skills and performance. They believe that sensory data need to be combined along with classification and characterization and correlation with physical–chemical data.

The following two studies come from Greece. The first one deals with the comparison and discrimination of two major monocultivar extra virgin olive oils from the two dominant olive cultivars in the southern region of Peloponnese, cv. Koroneiki and cv. Mastoides [5]. The fatty acid and sterolic profiles are used as compositional/traceability markers. This study aimed to evaluate the differences on specific chemical characteristics of the oils because of their botanical origin. Substantial compositional differences in the fatty acid and sterolic profiles between Koroneiki and Mastoides cultivars were detected by analysis of variance and principal component analysis.

The second study evaluated the extent to which Messinian olive oils comply with the "Kalamata Protected Designation of Origin (PDO)" regulation [6]. Quality indices were measured, and detailed analyses of sterols, triterpenic dialcohols, fatty acid composition, and wax content were conducted in a total of 71 samples of Messinian olive oils. Results demonstrated major fluctuations from the established EU regulatory limits on their chemical parameters. Results showed low concentrations of total sterols, high concentrations of campesterol, and a slight tendency towards high total erythrodiol content. Fatty acid composition and wax content were within the normal range.

The inter-varietal diversity of typical volatile and phenolic profiles of Croatian EVOO was investigated [7] to strengthen the varietal identities and position on the market of monovarietal and protected designation of origin (PDO) EVOO. 93 samples from six olive (*Olea europaea* L.) varieties were subjected to gas chromatography ion trap mass spectrometry (GC-IT-MS) and ultra-performance liquid chromatography with diode array detection (UPLC-DAD). Quantitative descriptive sensory analysis was also performed.

The last paper in this Special Issue deals with the quality of olive oil produced by different extractive processes carried out in a Tuscany oil mill at different harvesting periods in the same crop season, as affected by the presence of yeasts [8]. Yeast concentrations were higher in extraction processes at the end of the harvesting. Molecular methods were employed to identify twelve yeast species. Significant differences were shown by HS-SPME-GC-MS analysis of the volatile compounds in commercial EVOO inoculated with three yeast species (*Nakazawaea molendini-olei*, *Nakazawaea wickerhamii*, *Yamadazyma terventina*), and this was dependent on the strain inoculated. They also reported that the extraction plant might be colonized by some yeast species and this might affect the chemical and sensory characteristics of the EVOO.

The results of this Special Issue demonstrate the necessity for more stringent controls to increase quality and avoid fraud and adulteration of EVOO. In this direction, authentication methods are employed. It is quite important to safeguard quality and safety with various and promising tools that labs are equipped with, and sensory methodologies should be more critically combined with physico-chemical methods in order not only to comply with existing regulations, but also to provide consumers with a final product with added value. Companies should stop sacrificing value in the name of cost. Transparency procedures are essential in this direction.

Funding: This research received no external funding.

Conflicts of Interest: The author declares no conflict of interest.

References

1. Bubola, K.B.; Lukić, M.; Novoselić, A.; Krapac, M.; Lukić, I. Olive Fruit Refrigeration during Prolonged Storage Preserves the Quality of Virgin Olive Oil Extracted Therefrom. *Foods* **2020**, *9*, 1445. [CrossRef] [PubMed]
2. Violino, S.; Ortenzi, L.; Antonucci, F.; Pallottino, F.; Benincasa, C.; Figorilli, S.; Costa, C. An Artificial Intelligence Approach for Italian EVOO Origin Traceability through an Open Source IoT Spectrometer. *Foods* **2020**, *9*, 834. [CrossRef] [PubMed]
3. Aykas, D.P.; Karaman, A.D.; Keser, B.; Rodriguez-Saona, L. Non-Targeted Authentication Approach for Extra Virgin Olive Oil. *Foods* **2020**, *9*, 221. [CrossRef] [PubMed]
4. Barbieri, S.; Bubola, K.B.; Bendini, A.; Bučar-Miklavčič, M.; Lacoste, F.; Tibet, U.; Winkelmann, O.; García-González, D.L.; Toschi, T.G. Alignment and Proficiency of Virgin Olive Oil Sensory Panels: The OLEUM Approach. *Foods* **2020**, *9*, 355. [CrossRef] [PubMed]
5. Skiada, V.; Tsarouhas, P.; Varzakas, T. Comparison and Discrimination of Two Major Monocultivar Extra Virgin Olive Oils in the Southern Region of Peloponnese, According to Specific Compositional/Traceability Markers. *Foods* **2020**, *9*, 155. [CrossRef] [PubMed]
6. Skiada, V.; Tsarouhas, P.; Varzakas, T. Preliminary Study and Observation of "Kalamata PDO" Extra Virgin Olive Oil, in the Messinia Region, Southwest of Peloponnese (Greece). *Foods* **2019**, *8*, 610. [CrossRef] [PubMed]
7. Lukić, I.; Lukić, M.; Žanetić, M.; Krapac, M.; Godena, S.; Bubola, K.B. Inter-Varietal Diversity of Typical Volatile and Phenolic Profiles of Croatian Extra Virgin Olive Oils as Revealed by GC-IT-MS and UPLC-DAD Analysis. *Foods* **2019**, *8*, 565. [CrossRef] [PubMed]
8. Guerrini, S.; Mari, E.; Barbato, D.; Granchi, L. Mari Extra Virgin Olive Oil Quality as Affected by Yeast Species Occurring in the Extraction Process. *Foods* **2019**, *8*, 457. [CrossRef] [PubMed]

Article

Olive Fruit Refrigeration during Prolonged Storage Preserves the Quality of Virgin Olive Oil Extracted Therefrom

Karolina Brkić Bubola ***, Marina Lukić, Anja Novoselić, Marin Krapac and Igor Lukić**

Institute of Agriculture and Tourism, Karla Huguesa 8, 52440 Poreč, Croatia; marina@iptpo.hr (M.L.); anovoselic@iptpo.hr (A.N.); marin@iptpo.hr (M.K.); igor@iptpo.hr (I.L.)
* Correspondence: karolina@iptpo.hr; Tel.: +385-(0)52-408-341

Received: 11 September 2020; Accepted: 9 October 2020; Published: 12 October 2020

Abstract: With the aim to investigate the influence of post-harvest olive fruit storage temperatures on virgin olive oil production parameters, composition and quality, Istarska bjelica (IB) and Rosinjola (RO) fruits were stored for seven days at room temperature (RT), +4 °C and −20 °C prior to oil production. Lower temperatures delayed post-harvest maturation of IB fruits. Theoretical oil content did not change depending on the storage temperature, while the highest oil yield and extractability index were obtained after storage at RT. Chlorophylls decreased in IB-RT and in IB-20. A decrease in the sensory quality of oils was detected after fruit storage at RT and −20 °C, while the refrigeration temperature of +4 °C preserved it. Regarding the content of fatty acid ethyl esters, an increase was observed in IB-RT oils. Storage at RT increased the content of waxes, while the lower temperatures partially suppressed this phenomenon. In oils of both cultivars, storage at +4 °C preserved the concentration of most phenolic compounds at a level more similar to that of the fresh oil when compared to the other two treatments. In the production conditions, when prolonged fruit storage is necessary, refrigeration seems to be the most suitable option.

Keywords: olive fruits; storage temperature; virgin olive oil; FAEE; waxes; phenolic compounds; sensory analysis

---

1. Introduction

Olive oil production is a seasonal activity, and as such faces particular problems, which are more or less pronounced depending on the year-by-year situation, such as the global and micro climate conditions (rainfall and temperature), early or late ripening, decreased or surplus production (olive and oil yield) and even market positioning. Such factors affect harvesting decisions and may lead to inadequate/inefficient/insufficient oil production capacity in olive mills. Virgin olive oil (VOO) is obtained by several mechanical and physical processes, which begin with harvesting and post-harvest storage of olive fruits. The quality of VOO is closely related to the quality of olive fruits from which it is obtained [1]. In order to achieve high quality oil, it is recommended to process olives within 24 h after harvesting [2]. If processing is not done within that period, as for example when the capacity of available mills is insufficient, the fruits have to be stored for a certain period of time.

During a prolonged storage period, degenerative hydrolytic and oxidative processes in olive fruits start to develop. Under relatively high storage temperatures the proliferation of various microorganisms in olive fruit is accelerated, which often leads to many detrimental changes in olive physicochemical composition and the development of sensorial defects of fermentative origin, in the obtained olive oil [3,4].

Besides volatile compounds, principally associated with a decrease in olive oil sensory quality caused by prolonged olive fruit storage [4], other quality markers are affected as well. In olives with ruptured drupe (unhealthy, overripe, fractured during picking or storage), the fatty acids are liberated by the action of enzymes. In the presence of ethanol, also a product of fermentation, fatty acid ethyl esters (FAEE) are formed by esterification of free fatty acids with ethanol [5,6]. Recent studies suggest that the FAEE concentration is considered an indicator of olive oil quality [7,8]. Higher storage temperatures, such as the ambient temperature of 25 °C, accelerate the processes of olive ripening [9]. As olive fruit ripens, its exocarp becomes thinner, and fruit tissues become softer [10]. Along with these changes, particular compounds are being extracted into oil, such as waxes from the waxy surface layer of the cuticle of olive fruit. The content of waxes in the oil can also be considered a quality indicator [6] although it is officially used as a marker for distinguishing olive from olive-pomace oil [8,11]. Phenolic compounds, in addition to their involvement in VOO taste characteristics, have antioxidant properties and, therefore, significantly contribute to the oxidative stability and health benefits associated with VOO [12]. Their presence in oil depends on the interaction between genetic (cultivar), environmental (cultivation, harvest and postharvest conditions) and physiological factors (fruit ripening degree and sanitary conditions), and processing conditions [13,14]. Previous studies revealed that phenolic compounds are strongly affected by olive fruit storage conditions, especially secoiridoids, whose concentrations decrease with longer storage times and higher storage temperatures [15,16]. Several authors have found that, due to genetically predetermined enzymatic activity, the polyphenol content of obtained VOO behaved differently among cultivars even if the fruits were stored under the same conditions prior to oil production [16,17]. Considering all the reasons mentioned above, optimal duration and temperature of fruit storage are obviously very important factors for obtaining high quality VOO in general [9] and are also crucial for preserving particular sensory profiles typical for VOO of certain cultivars, consisting of high levels of phenolic compounds and related high intensities of bitterness and pungency, as in the case of Croatian autochthonous Istarska bjelica and Rosinjola [18,19].

Currently, the main strategies for avoiding deterioration of olive fruit during storage and production of lower quality VOO include reducing the period between harvesting and processing by increasing the production capacity of olive mills, as well as applying conditions aimed to ensure lesser contact between fruits (larger storage spaces, perforated boxes, etc.) [20]. Storage of olive fruits at lower temperatures was also recognized as an alternative that could allow more flexibility during harvesting and oil production [17]. In general, low storage temperatures decrease water activity and inhibit microbial growth [21], slow down enzymatic and biochemical reactions [22], decrease fruit respiration and delay harvested fruit maturation [10,23]. On the other hand, low temperatures may cause physiological damage of the fruit [24] since ice crystals formed during frozen storage may break down the cell structure and at the same time allow the contact between various enzymes and substrates that could affect VOO quality [25,26]. Very low temperatures during olive fruits development and harvest, with related cycles of freezing and defrosting of fruits on the tree, are known to cause biochemical changes, which significantly modify its volatile compound and phenolic profiles [22] and could induce the development of the "frostbitten olives" sensory defect in olive oil [27].

Several studies have been conducted in order to investigate appropriate olive fruit storage conditions, including mostly ambient and refrigeration temperatures [9,16,21,23,28]. A few studies that examined the effects of frozen storage included either a very short storage period of one day [29], which might not be sufficient in practical conditions when delays in olive processing are larger, or a very long storage period of 6 months [26], which is not applicable from a practical point of view. These studies focused mostly on the basic quality parameters and phenolic and volatile profiles of the obtained oils and, to our knowledge, none of them investigated the influence of low storage temperatures on the contents of FAEE and waxes.

The aim of this study was to investigate the influence of low temperature during prolonged post-harvest storage of olive fruit on the production parameters (oil content in the fruits, oil yield and extractability index), composition (the content of pigments, FAEE, waxes and phenolic compounds) and sensory quality of olive oil obtained therefrom. The experiment was performed with olive fruits of two autochthonous Croatian cultivars, Istarska bjelica and Rosinjola, non-stored (control) and stored for seven days at three different temperatures, including room temperature, refrigerating at +4 °C and freezing at −20 °C. The main intention was to evaluate the possibility of prolonging the storage period of olive fruits to a reasonable time still feasible in practice (one week), without compromising the chemical and sensory quality of VOO, which would possibly contribute to overcoming the problems with olive mill overloading during the harvesting period. In addition, particular attention was devoted to the influence of olive fruit storage temperature on both positive and negative sensory characteristics of the obtained VOO, in order to deepen the understanding of the origin of some VOO defects, such as "frostbitten olives", otherwise associated with the impact of low ambient temperatures during olive fruit maturation.

2. Materials and Methods

2.1. Olive Fruits, Storage Treatments and Virgin Olive Oil Production

Healthy olive fruits from Istarska bjelica (IB) and Rosinjola (RO), two Croatian autochthonous olive cultivars (*Olea europaea* L.), were manually harvested at the end of October 2016 and the beginning of November 2016, respectively. Both cultivars were grown in the same experimental olive orchard of the Institute of Agriculture and Tourism (Poreč, Croatia). The ripening index of olive fruits was determined by the protocol described by Beltran et al. [30].

Olive fruits were divided in twelve batches of 3 kg per cultivar. Three batches per cultivar were processed into oil immediately after harvest (control oil), and the rest of the olive fruits was stored for seven days at three different temperatures prior to olive oil production: at room temperature of 22 ± 4 °C (RT), at +4 °C in a refrigerator and at −20 °C in a freezer. The olive fruits stored at 4 °C and −20 °C were allowed to reach room temperature before milling, which lasted about 2 and 6 h, respectively. Fruits were crushed by a hammer crusher and olive paste was malaxed for 45 min at 25 °C using vertical thermostated olive paste mixers. Olive oil extraction was done by a laboratory centrifuge (Abencor, MC2 Ingeneria y Sistemas, Seville, Spain). In order to obtain enough oil for analysis, fruits representing one batch of 3 kg, were processed in triplicates and obtained oils were mixed in one oil sample. Obtained oil samples ($n = 12$ per cultivar, 3 control oils and 3 samples per each storage time/temperature) were left to sediment naturally for ten days and were then decanted. The analyses of oil samples started immediately after decantation. Samples were stored in non-transparent bottles at 16–18 °C during the time of analysis.

2.2. Oil Content, Oil Yield and Extractability Index

Theoretical oil content in the fruit (expressed on fresh and on dry weight based on the gravimetric determination of water in fruit) was determined from the olive paste obtained after crushing using Soxtec Avanti 2055 apparatus (Foss Tecator, Höganäs, Sweden) according to the method described by Brkić et al. [31].

Oil yield (%) was calculated from three parallel processing repetitions, multiplying by 100 the mass ratio of mechanically extracted oil (g) and centrifuged olive paste (g) [32].

Olive oil extractability index (EI) was calculated according to Beltran et al. [30] using the formula: $EI = V \times d/W \times F \times 100$, where V (mL) is a volume of olive oil extracted, d (0.915 g/mL) is the average olive oil density, W (g) is olive paste weight and F (%) is the oil content of the fruit (on fresh weight).

2.3. Analysis of VOOs Pigments

Chlorophyll and carotenoid concentrations were determined using a Varian Cary 50 UV/Vis spectrophotometer (Varian Inc., Harbour City, CA, USA) following the procedure of Mínguez-Mosquera et al. [33] and expressed as pheophytin a and lutein content, respectively.

2.4. Sensory Analysis

Quantitative descriptive sensory analysis of VOO samples was performed by the Panel for sensory assessment of VOO, accredited for VOO sensory analysis according to the EN ISO/IEC 17025:2007 and recognized in continuation by the International Olive Council (IOC) from 2014. The panel consisted of eight assessors (5 female, 3 male, average age 35) trained for VOO sensory analysis according to the IOC method [34].

2.5. FAEE and Waxes

FAEE and waxes were determined by the IOC method [11] employing extraction by column chromatography and analysis by gas chromatography (GC) with flame-ionization detection using a Varian 3350 gas chromatograph (Varian Inc., Harbour City, CA, USA).

2.6. Analysis of Phenolic Compounds

Extraction and HPLC analysis of phenolic compounds using an Agilent Infinity 1260 System (Agilent Technologies, Santa Clara, CA, USA) in oil samples was performed according to the method proposed by Jerman Klen et al. [35] and slightly modified by Lukić et al. [36].

Identification of peaks was performed by comparing retention times and UV/Vis spectra with those of pure standards and those from the literature [35]. The detection was carried out at 280 nm for simple phenols, lignans, secoiridoids and vanillic acid, at 320 nm for vanillin and *p*-coumaric acid, and at 365 nm for flavonoids. For quantification, standard calibration curves were made for tyrosol, hydroxytyrosol, vanillic acid, vanillin, *p*-coumaric acid, luteolin, apigenin, pinoresinol and oleuropein. Based on constructed calibration curves, concentrations of samples were expressed as mg/kg oil. Semiquantitative analysis was performed for hydroxytyrosol acetate, acetoxypinoresinol and secoiridoids, where the concentration was expressed as hydroxytyrosol, pinoresinol and oleuropein, respectively, assuming a response factor equal to one. Total phenolic content was presented as the sum of all the identified phenolic compounds.

2.7. Data Elaboration

To investigate the effects of different fruit storage temperature on the VOO's investigated parameters, results of the chemical and sensorial analysis were subjected to a one-way analysis of variance (ANOVA). Means were compared by the Tukey's honest significant difference test at the level of $p < 0.05$. Statistical analysis was carried out using Statistica v. 13.2 software (Stat-Soft Inc., Tulsa, OK, USA).

3. Results and Discussion

3.1. Oil Content, Extractability Index and Ripening Index

In order to monitor the accumulation of oil in olive fruits during storage, oil content on dry matter was determined (Table 1). It was determined that it did not change significantly depending on the fruit storage temperature in the case of both investigated cultivars. This result indicates that the accumulation of oil did not continue during fruit storage, which is in agreement with the findings of Inarejos-García et al. [37] during the storage of Cornicabra cultivar fruits at 10 °C and 20 °C for three weeks, and that of Yousfi et al. [23] in the case of Arbequina olives stored up to three weeks at 3 °C and 18 °C.

Table 1. Ripening index (RI), oil on dry weight, yield and extractability index (EI) of Istarska bjelica (IB) and Rosinjola (RO) cultivar fresh fruits immediately after harvest (control) and fruits stored seven days at three different storage temperature (RT—room temperature, +4 °C and −20 °C) prior to production.

	RI	% Oil on Dry Weight	Yield (%)	EI
IB-control	1.02 ± 0.10 [b]	40.30 ± 1.60	10.03 ± 0.20 [b]	0.45 ± 0.02 [b]
IB-RT	1.73 ± 0.07 [a]	37.95 ± 3.77	11.37 ± 0.38 [a]	0.55 ± 0.04 [a]
IB+4	1.11 ± 0.05 [b]	40.84 ± 1.02	9.88 ± 0.17 [b]	0.44 ± 0.00 [b]
IB-20	1.16 ± 0.10 [b]	40.90 ± 2.57	9.19 ± 0.42 [b]	0.41 ± 0.03 [b]
RO-control	1.64 ± 0.07	35.53 ± 2.76	5.11 ± 0.21 [b]	0.22 ± 0.01 [b]
RO-RT	1.75 ± 0.05	38.63 ± 4.51	7.14 ± 0.15 [a]	0.29 ± 0.03 [a]
RO+4	1.72 ± 0.06	35.75 ± 1.95	5.25 ± 0.21 [b]	0.23 ± 0.02 [b]
RO-20	1.58 ± 0.11	39.37 ± 3.00	5.33 ± 0.16 [b]	0.21 ± 0.02 [b]

Results are expressed as mean values ± standard deviation of three technical repetitions. Mean values labeled with a different superscript letter, within the same column and same cultivar are statistically different (Tukey's test, $p < 0.05$). In case there were no statistically significant differences the letters were omitted.

On the other hand, considering the processing parameters, olive oil yield and extractability index (EI), the highest values were obtained in the case of storage at RT (Table 1). Yousfi et al. [10] reported that olives stored under ambient temperature (18 °C) exhibited higher respiration rates than refrigerated ones, which is associated with fruit ripening and, consequently, softening. As a consequence of ripening, degradation of walls of oil-bearing cells is facilitated and the extraction process is improved, which could have been the cause of the increase of the olive oil yield and EI in the RT stored fruits in this study. In IB fruits, a significant increase in RI was observed after seven days of storage and it depended on the temperature, since it increased only in the case of fruits stored at RT. In RO fruits significant differences between the treatments were not found (Table 1). García et al. [38] have also found that cold storage (5 °C) could delay ripening of Blanqueta and Villalonga olives compared to storage at ambient temperature (12 ± 5 °C). Different from the results of this study, the extractability of Arbequina olives stored up to 21 days at 3 °C and 18 °C showed a similar oil yield to the initial unstored sample [23]. Extractability index is highly dependent on the cultivar and its fruits properties [30]. Both of the investigated autochthonous Croatian cultivars had the value of EI in line with most of the leading Spanish olive cultivars [30], indicating their good potential for oil production regardless of the storage temperature of the fruit prior to processing.

3.2. VOO Pigments

Considering the chlorophyll content (Figure 1) in the VOO obtained from IB fruits, similar content was determined in IB+4 as in IB-control oil, while a mild decrease in IB-RT and a pronounced decrease in IB-20 compared to IB-control oil was determined. García et al. [39] have found that the maturation of Picual cultivar olive fruits was delayed while stored at 5 °C or 8 °C, compared to oils obtained from fruits stored at ambient temperature. The cause of this was low temperature, which delayed the destruction of chlorophyll pigments and their substitution by anthocyanins in the cells of olive skin during fruit maturation [39]. On the other hand, Morelló et al. [40] have found a decrease in the content of pigments (chlorophylls and carotenoids) in Arbequina oils obtained from fruits that were frozen on the trees seemingly due to the activity of chlorophyllase enzymes involved in the loss process. By visual inspection it was observed that the chilling injuries of IB-20 fruits occurred in the form of browning, which probably influenced a decrease of chlorophylls in the obtained oils. Chlorophyll content was also low in the oil obtained from Koroneiki olives after 30 days of storage at 0 °C, probably due to chilling [41]. The oil from olives stored at 5 °C had slightly lower chlorophyll content, while the oil from olives stored at 7.5 °C had similar chlorophyll content as the oil from freshly harvested olives [41]. A significant effect of different storage temperature on the chlorophyll content was not observed in RO oils, probably because its fruits did not continue to ripen during storage (Table 1), while fruit injuries as a result of freezing during storage were not observed by visual inspection. Yousfi et al. [10] also found that storage conditions (3 °C and 18 °C during 3 weeks) did not affect the content of chlorophylls in Arbequina fruits.

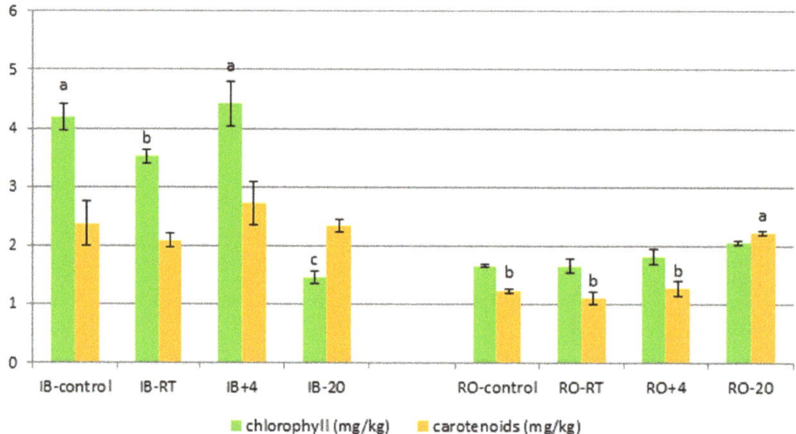

Figure 1. Concentrations of chlorophyll and carotenoids (mg/kg) in Istarska bjelica (IB) and Rosinjola (RO) monovarietal virgin olive oils obtained from fresh fruits immediately after harvest (control) and oils obtained from fruits stored seven days at three different storage temperature (RT—room temperature, +4 °C and −20 °C) prior to production. Results are expressed as mean values ± standard deviation of three technical repetitions. Mean values labeled with a different letter, within one parameter and one cultivar are statistically different (Tukey's test, $p < 0.05$). In case there were no statistically significant differences the letters were omitted.

Carotenoids content was not significantly changed depending on the fruit storage temperature in both IB and RO oils, except in the case of RO-20 oil where an increase was detected when compared to RO-control oil (Figure 1). Yousfi et al. [10] have found that carotenoids were not affected by the storage conditions (3 °C and 18 °C during 3 weeks) applied to the Arbequina fruits. The increased content of carotenoids in RO-20 could be related to a decrease in the consistency of the chloroplast wall caused by low storage temperature that facilitates the release of these pigments into olive oil [23].

3.3. Sensory Quality

After fruit storage, panelists observed a decrease in fruitiness and a major positive aroma sensory characteristic in the oil samples of both cultivars, (Figure 2). The highest decrease was determined in oils stored at −20 °C (approximately 3 intensity units compared to control oils). The taste characteristics of the oils, such as bitterness and pungency, were less altered after the prolonged fruit storage than the olfactory characteristics, although in most cases slightly lower intensities were determined compared to the control, except for IB+4 oil, which was similar to IB-control oil (Figure 2). García et al. [39] found that bitterness and sensory quality of Picual oils obtained from fruits stored at RT decreased rapidly and that the loss was slowed down during storage at 5 °C. Morelló et al. [40] found that Arbequina olive oil had a decreased intensity of bitterness and pungency when produced from fruits that have been frozen on the trees. Inarejos-García et al. [37] observed a larger reduction of bitterness, determined as K225, in Cornicabra olive oil produced from fruits stored for 5 days at 20 °C compared to that obtained from olives stored at 10 °C for a week. The same authors concluded that prolonged storage could be useful for modifying the taste of oils of phenol-rich cultivars, such as Spanish Cornicabra, characterized by intense bitter taste that could affect consumers' preferences. On the other hand, preserving the bitterness and pungency in IB and RO oils could be very important, since the mentioned sensorial characteristics were shown to be typical for these autochthonous monovarietal olive oils [18,19], especially because these cultivars are included in the production of Croatian oils under the protected denomination of origin (PDO) "Istra", which gives them an added value.

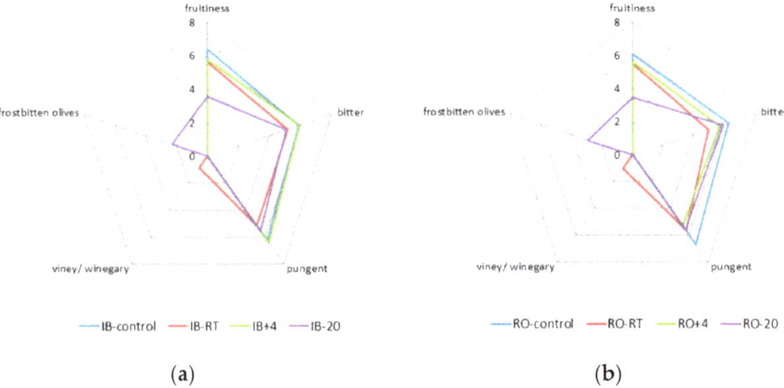

(a)　　　　　　　　　　　　(b)

Figure 2. Results of sensory analysis of (**a**) Istarska bjelica (IB) and (**b**) Rosinjola (RO) monovarietal virgin olive oils obtained from fresh fruits immediately after harvest (control) and oils obtained from fruits stored seven days at three different storage temperature (RT—room temperature, +4 °C and −20 °C) prior to production. Results are expressed as mean values of the medians of three technical repetitions.

In the oil samples obtained from the fruits stored for seven days at RT and −20 °C negative sensory characteristics were determined (Figure 2). In RT oils of both cultivars a slight intensity (around 1) of the "viney/winegary" defect was noted, while the defect "frostbitten olives" was recognized as the main defect in the oils obtained from fruits stored at −20 °C, with the intensity of 2.3 in the case of IB and 2.9 in the case of RO cultivar oil respectively. IB and RO oils obtained from the fruits stored at RT and −20 °C could not be classified as extra virgin olive oils (the highest quality category), since, according to the intensity of the recognized defects, they belonged to the virgin olive oil category (EEC, 1991). Sensory defects were prevented by the storage at +4 °C (Figure 2), indicating that +4 °C was the most appropriate temperature for fruit storage in order to assure good sensorial quality of the obtained oils. "Viney/winegary" defect, and "fusty", "muddy sediment" and "musty" defects usually develop in oils because of the proliferation of particular microorganisms (lactic, acetic and enteric bacteria, fungi and *Pseudomonas*) on olive fruits during unsuitable storage conditions [3,4]. Kiritsakis et al. [41] reported that Koroneiki olives stored at 0 °C and 5 °C had no fungus development, while this was not the case at 7.5 °C, where the noticed increase in oil acidity was a result of fungal lipase activity [41], which can cause development of sensory defects. Garcia et al. [38] have found a different response of the sensory quality of different cultivars: Blanqueta olive oil developed defects more rapidly than Villalonga olive oil during 30 days of storage at ambient temperature and at 5 °C, and the development of off-flavors was more rapid at ambient than at low temperature.

Freeze injuries are a consequence of olive fruit cell dehydration and destruction caused by ice crystals forming inside the parenchyma cells, which cause destruction of cell membranes, leading to cell death and high oxidation of cell contents [40]. This is the consequence of the contact between enzymes and their respective substrates, which may have an effect on the composition of the obtained olive oil [40]. Freeze injuries were not detected on olives of both cultivars stored at +4 °C, which is in agreement with the result for Koroneiki olives stored at 5 °C and 7 °C for 40 days [41]. On the other hand, freeze injuries in the form of fruit skin browning and shriveling were detected by visual inspection on olives stored at −20 °C, which finally resulted in olive oils with perceived "frostbitten olives" defect (Figure 2). Although some authors reported that severe freezing damage of olive fruits on trees during winter time could have negative influence on the sensory characteristics of olive oil [42,43], there is little information on how controlled freezing temperatures during olive fruit storage influence the sensorial profile of obtained oils. Romero et al. [43] reported two different descriptions of the "frostbitten olives" defect, which depend on whether the temperature changes took place abruptly, with rapid freeze–thaw cycles, or gradually. They reported that oils were grouped based on the concentrations of volatile compounds into two clusters, characterized by different profiles. The first was characterized by descriptors such as "soapy" and "strawberry-like" and the characteristic presence of ethyl 2-methylbutanoate and ethyl propanoate, and the second by "wood" and "humidity" descriptors and high concentrations of pentanal and octanal. In this study, the sensory profile of the "frostbitten olives" defect perceived by the panelists was described using a descriptor "wet wood" (Figure 2), which was more similar to the second profile reported by Romero et al. [43], indicating that a gradual drop of temperature took place during the controlled freezing at −20 °C, with the formation of extracellular ice and evaporation of liquid water inside the cells. According to Romero et al. [43], as water is removed from the cells, ice continues to grow and damages the cells until they break down.

3.4. FAEE and Waxes

FAEEs are closely related to health conditions of the fruits and their concentration is higher in olives that underwent hydrolytic and fermentative processes that produce additional amounts of free fatty acids and alcohols [44]. Regarding the FAEE parameter, there were no significant differences among the treatments in RO oils. However, an increase in FAEE concentration was observed in IB-RT compared to the IB-control oil (Table 2), which was probably a result of the softening and damage of the fruit tissue during prolonged storage as a consequence of accelerated ripening of the fruits at higher storage temperature (Table 1). Jabeur et al. [44] have found an increase in FAEE concentration during Chemlali olive fruit storage at ambient temperature (12–18 °C) for 25 days in closed plastic bags and in open perforated plastic boxes, probably a consequence of microorganism fermentation activity. In the oil samples investigated in this study, total FAEE concentration ranged from 4 to 12 ppm and as such was below the maximum legal limit of ≤35 ppm set for EVOO [8]. Although the FAEE values did not surpass the maximum legal limit, they were in line with the results obtained by sensory analysis of the IB-RT oil, where a slight intensity of "viney/winegary" defect was determined (Figure 2). The correlation found between FAEE amounts and fermentative defects was probably due to their common origin [1,6]. On the other side, the intensities of non-fermentative defects, e.g., "frostbitten olives", determined in the oils obtained after frozen storage of the fruits of both cultivars (Figure 2), are not related to the concentrations of FAEE as reported by the literature [1].

The concentration of waxes (C246) in the investigated samples ranged from 15 to 50 ppm (Table 2). Although the obtained values did not surpass the maximum legal limit for EVOO of ≤150 ppm [8], the RT treatment showed a significant increase in the concentration of most waxes compared to the controls and the other two treatments in the oils from both cultivars. Storage at room temperatures may cause acceleration of fruit ripening [9], which is followed by fruit cuticle thinning and softening of fruit tissue [10]. As a consequence of those changes, waxes from the waxy surface layer of the cuticle of olive fruit could be more easily extracted into oil. The more mature, and possibly the more degraded olive fruits were (as in the case of IB fruits stored at RT, Table 1), the higher was the amount of waxes extracted, which supported the assertion that higher concentration of waxes could indicate lower quality of olive oil [6,44]. The storage of fruits at temperatures lower than RT resulted in lower concentration of waxes in the obtained oils (Table 2), probably due to the delay in fruit ripening.

Table 2. Concentrations of ethyl esters and waxes (mg/kg) in Istarska bjelica (IB) and Rosinjola (RO) monovarietal virgin olive oils obtained from fresh fruits immediately after harvest (control) and oils obtained from fruits stored seven days at three different storage temperature (RT—room temperature, +4 °C and −20 °C) prior to production.

	Ethyl Esters			Waxes					
	EE C16	EE C18	FAEE	C40	C42	C44	C46	C246	C0246
IB-control	2.58 ± 1.53	1.95 ± 0.87 [c]	4.53 ± 1.8 [b]	12.59 ± 0.76 [c]	9.54 ± 0.98 [c]	2.59 ± 0.22 [b]	2.92 ± 0.72	15.06 ± 1.28 [c]	27.64 ± 1.82 [c]
IB-RT	3.48 ± 1.04	8.24 ± 0.82 [a]	11.71 ± 0.28 [a]	21.13 ± 1.66 [a]	15.08 ± 0.95 [a]	3.42 ± 0.15 [a]	3.29 ± 0.25	21.79 ± 0.71 [a]	42.92 ± 2.37 [a]
IB+4	2.71 ± 0.53	4.38 ± 1.14 [b]	7.09 ± 1.61 [b]	17.28 ± 0.42 [b]	10.48 ± 0.27 [bc]	2.67 ± 0.11 [b]	2.37 ± 0.25	15.52 ± 0.39 [c]	32.81 ± 0.80 [b]
IB-20	2.64 ± 0.12	3.56 ± 0.68 [bc]	6.21 ± 0.56 [b]	16.38 ± 1.26 [b]	12.02 ± 0.80 [b]	3.79 ± 0.49 [a]	2.75 ± 0.31	18.56 ± 1.14 [b]	34.94 ± 2.03 [b]
RO-control	5.73 ± 1.10	2.44 ± 1.64	8.17 ± 2.74	14.59 ± 0.39	24.25 ± 6.67 [ab]	5.79 ± 1.15 [b]	1.46 ± 0.35 [b]	31.49 ± 7.47 [b]	46.08 ± 7.08 [b]
RO-RT	3.18 ± 1.77	3.29 ± 1.68	6.47 ± 2.22	19.17 ± 2.44	34.21 ± 3.38 [a]	12.59 ± 0.85 [a]	3.85 ± 1.13 [a]	50.65 ± 5.07 [a]	69.82 ± 7.48 [a]
RO+4	2.88 ± 1.08	2.61 ± 1.88	5.49 ± 2.84	14.06 ± 2.47	22.84 ± 6.69 [b]	8.42 ± 3.22 [b]	1.80 ± 0.64 [b]	33.06 ± 10.53 [b]	47.12 ± 12.87 [b]
RO-20	2.96 ± 0.58	1.78 ± 0.93	4.73 ± 1.49	14.42 ± 2.47	20.44 ± 0.72 [b]	6.62 ± 0.26 [b]	1.79 ± 0.20 [b]	28.85 ± 1.00 [b]	43.28 ± 2.71 [b]

Results are expressed as mean values ± standard deviation of three technical repetitions. Mean values labeled with a different superscript letter, within the same column and same cultivar are statistically different (Tukey's test, $p < 0.05$). In case there were no statistically significant differences the letters were omitted. EE C16—ethyl palmitate, EE C18—ethyl stearate, FAEE—fatty acid ethyl esters. C246 = C42 + C44 + C46, C0246 = C40 + C42 + C44 + C46.

3.5. Phenolic Compounds

A reduction of the total concentration of the identified phenolic compounds was detected in the oils from particular treatments, but the highest decrease compared to the control treatment was detected in the oils obtained from the fruits of both cultivars stored at −20 °C (Table 3). Hachicha Hbaieb et al. [9] have found that negative effects of storage time on phenolic compounds in oils were enhanced by an increase in storage temperature from 4 to 25 °C. Yousfi et al. [10] found that the main phenolic compounds in VOO exhibited a reduction during 15 days of fruit storage, which was in correlation with the increase in the applied temperature (from 2 to 18 °C). Other authors, who investigated the influence of freezing of fruits on trees, reported a decrease in the concentration of phenolic compounds in the obtained oils. They explained it as a consequence of fruit freeze injuries, which lead to cell dehydration and destruction of cell membranes, and consequently to cell death and high oxidation of cell contents as a result of the contact between enzymes and their respective substrates, which might have affected the phenolic composition of the oils [40,42]. Morelló et al. [40] investigated the influence of freezing of Arbequina fruits on trees on phenols in the obtained oils. They have found that total phenols and secoiridoids decreased after frost because ice crystals destructed olive tissues, which encouraged the oxidative degradation of phenolic compounds in reactions catalyzed by polyphenol oxidase enzyme [40]. Masella et al. [26] investigated the difference between three different methods of freezing of olives and found a significant reduction of total phenols in oils obtained from fruits after 6 months of storage at freezing temperatures (about 40% of the control oils) regardless of the freezing method used. It must be mentioned that not all the identified phenolic compounds absorb UV light equally, meaning the use of oleuropein as a standard for all secoiridoids with the response factor equal to one in the HPLC-DAD analysis in this study might have resulted with an overestimation of the reduction of the total phenol concentration in the oils of particular treatments. For example, p-HPEA-EDA (oleocanthal) has a lower response factor in comparison to 3,4-DHPEA-EDA (oleacein), and the same applies for the corresponding aglycone isomers of ligstroside and oleuropein. This difference is related to the different substitution of the aromatic ring. The underestimation of the secoiridoids bearing the tyrosol moiety might have had a notable impact on the calculated total phenol concentrations. More specifically, although the reduction noted is relative for each compound, the actual total phenolic loss might be less than reported.

Considering the secoiridoid group, a reduction was found in the case of treatments at RT and −20 °C in the oils from both cultivars. Since secoiridoid compounds are strongly related to the VOO shelf life [45], it can be assumed that the oils stored at +4 °C would have the longest shelf life among the oils obtained from the stored fruits. Reduction of secoiridoids was lower in RO oils, which initially had a lower concentration of total secoiridoids compared to IB oils. Li et al. [15] noticed that the higher the initial concentration of these phenolic compounds in oil, the faster they decrease during storage, possibly because higher concentrations are more susceptible to oxidation with respect to other antioxidants in olive oil. Guillaume et al. [42] noted a reduction of the concentration of secoiridoids in the oils obtained from the frost-damaged fruits of three olive cultivars (Frantoio, Barnea and Picual) grown in Australia. Hachicha Hbaieb et al. [9] have also observed a larger decrease in secoiridoids concentration in Arbequina and Chétoui oils obtained from fruits stored at 25 °C than at 4 °C, and related this to the lower β-glucosidase activity determined in olive fruits from the former treatment.

Table 3. Concentration of phenolic compounds in Istarska bjelica (IB) and Rosinjola (RO) monovarietal virgin olive oils obtained from fresh fruits immediately after harvest (control) and oils obtained from fruits stored seven days at three different storage temperature (RT—room temperature, +4 °C and −20 °C) prior to production.

Phenolic Compounds (mg/kg)	IB-Control	IB-RT	IB+4	IB-20	RO-Control	RO-RT	RO+4	RO-20
Simple phenols								
hydroxytyrosol	9.30 ± 0.96 [b]	27.50 ± 6.48 [a]	20.15 ± 0.90 [a]	7.31 ± 2.83 [b]	10.76 ± 1.51 [b]	31.47 ± 2.62 [a]	22.25 ± 5.72 [a]	20.93 ± 5.40 [ab]
tyrosol	7.67 ± 0.73 [c]	40.16 ± 8.53 [a]	20.22 ± 1.62 [b]	11.42 ± 2.36 [bc]	2.13 ± 0.26 [b]	9.12 ± 3.53 [a]	8.16 ± 2.07 [ab]	3.62 ± 3.09 [ab]
hydroxytyrosol acetate	1.14 ± 0.39 [b]	2.86 ± 0.60 [a]	2.01 ± 0.18 [ab]	1.11 ± 0.27 [b]	0.87 ± 0.24	1.31 ± 0.22	1.20 ± 0.09	0.85 ± 0.14
vanillin	0.20 ± 0.02	0.19 ± 0.02	0.21 ± 0.01	0.20 ± 0.03	0.31 ± 0.04	0.25 ± 0.07	0.17 ± 0.06	0.20 ± 0.09
Total simple phenols	18.30 ± 0.88 [c]	70.72 ± 15.19 [a]	42.58 ± 0.53 [b]	20.05 ± 5.36 [c]	14.08 ± 2.03 [b]	42.15 ± 5.87 [a]	31.78 ± 7.82 [a]	25.60 ± 8.27 [ab]
Secoiridoids								
secologanoside	0.34 ± 0.09	0.21 ± 0.21	0.15 ± 0.18	0.24 ± 0.13	0.13 ± 0.14	0.18 ± 0.19	0.17 ± 0.08	0.26 ± 0.24
elenolic acid glucoside (isomer)	1.99 ± 0.46 [a]	0.74 ± 1.13 [ab]	0.44 ± 0.62 [ab]	0.16 ± 0.10 [b]	0.43 ± 0.67	0.73 ± 1.17	1.28 ± 1.07	0.47 ± 0.70
3,4-DHPEA-EDA	120.02 ± 36.92 [a]	29.58 ± 9.90 [c]	84.78 ± 3.87 [ab]	50.36 ± 4.75 [bc]	73.30 ± 5.57 [a]	48.63 ± 10.97 [a]	58.65 ± 14.35 [a]	35.30 ± 5.75 [b]
oleuropein aglycone (isomer I)	62.34 ± 17.93	46.23 ± 3.49	52.90 ± 1.78	44.50 ± 9.08	152.05 ± 6.27 [a]	98.64 ± 16.17 [bc]	111.01 ± 7.86 [bc]	73.26 ± 11.66 [c]
p-HPEA-EDA	49.03 ± 8.82 [a]	31.47 ± 5.70 [b]	47.18 ± 2.80 [a]	35.86 ± 1.82 [ab]	11.24 ± 0.66 [a]	10.13 ± 1.65 [ab]	9.85 ± 2.20 [ab]	6.79 ± 0.54 [b]
oleuropein + ligstroside aglycones I and II	51.52 ± 4.84 [a]	28.75 ± 5.73 [b]	35.64 ± 2.18 [b]	33.31 ± 3.16 [b]	14.55 ± 0.60 [a]	10.27 ± 0.99 [bc]	12.05 ± 1.81 [ab]	8.02 ± 1.61 [c]
oleuropein aglycone (isomer II)	91.19 ± 15.37 [a]	60.43 ± 2.27 [b]	83.55 ± 4.17 [ab]	69.99 ± 8.19 [ab]	68.50 ± 2.39	72.68 ± 8.09	59.62 ± 5.95	62.16 ± 0.99
ligstroside aglycon (isomer III)	1.61 ± 0.21 [b]	2.72 ± 0.78 [ab]	3.59 ± 0.98 [a]	4.11 ± 0.05 [a]	1.41 ± 0.14 [a]	1.52 ± 0.04 [a]	1.53 ± 0.37 [a]	0.80 ± 0.09 [b]
oleuropein aglycone (isomer III)	9.93 ± 2.77	11.37 ± 1.23	14.87 ± 2.44	15.87 ± 3.04	6.33 ± 0.36	7.82 ± 1.00	7.20 ± 1.59	5.32 ± 0.49
Total secoiridoids	387.98 ± 71.14 [a]	211.49 ± 22.77 [c]	323.09 ± 13.86 [ab]	254.41 ± 21.81 [bc]	327.94 ± 8.77 [a]	250.60 ± 33.36 [b]	261.35 ± 35.28 [ab]	192.38 ± 19.30 [b]
Lignans								
pinoresinol	12.46 ± 2.11 [b]	17.35 ± 0.37 [a]	17.94 ± 0.06 [a]	10.96 ± 0.40 [b]	3.11 ± 0.11 [a]	2.96 ± 0.04 [a]	3.03 ± 0.02 [a]	2.72 ± 0.08 [b]
acetoxypinoresinol	24.85 ± 1.60 [a]	24.84 ± 0.51 [a]	25.80 ± 1.01 [a]	20.99 ± 1.08 [b]	30.25 ± 0.55 [a]	30.66 ± 0.78 [a]	30.59 ± 0.95 [a]	25.53 ± 1.64 [b]
Total lignans	37.31 ± 3.56 [b]	42.20 ± 0.89 [ab]	43.75 ± 1.08 [a]	31.96 ± 0.72 [c]	33.36 ± 0.66 [a]	33.62 ± 0.75 [a]	33.62 ± 0.93 [a]	28.25 ± 1.60 [b]
Phenolic acids								
vanillic acid	0.96 ± 0.09 [b]	0.86 ± 0.08 [b]	0.94 ± 0.01 [b]	1.20 ± 0.04 [a]	1.03 ± 0.03	1.05 ± 0.06	1.06 ± 0.10	1.03 ± 0.73
p-coumaric acid	1.74 ± 0.21 [b]	5.16 ± 1.17 [a]	2.12 ± 0.11 [b]	1.28 ± 0.05 [b]	0.46 ± 0.01 [c]	4.74 ± 0.26 [a]	1.26 ± 0.06 [b]	0.43 ± 0.10 [c]
Total phenolic acids	2.70 ± 0.27 [b]	6.02 ± 1.11 [a]	3.06 ± 0.10 [b]	2.48 ± 0.05 [b]	1.49 ± 0.04 [b]	5.79 ± 0.32 [a]	2.33 ± 0.16 [b]	1.46 ± 0.71 [b]
Flavonoids								
luteolin	1.96 ± 0.25 [a]	1.14 ± 0.05 [b]	1.43 ± 0.10 [b]	0.70 ± 0.05 [c]	0.96 ± 0.11 [a]	0.77 ± 0.04 [a]	0.90 ± 0.08 [a]	0.49 ± 0.09 [b]
apigenin	1.01 ± 0.09 [a]	0.60 ± 0.01 [bc]	0.72 ± 0.05 [b]	0.48 ± 0.02 [c]	0.35 ± 0.04 [a]	0.29 ± 0.02 [b]	0.34 ± 0.01 [a]	0.24 ± 0.04 [b]
Total flavonoids	2.97 ± 0.33 [a]	1.74 ± 0.07 [b]	2.15 ± 0.16 [b]	1.18 ± 0.05 [c]	1.32 ± 0.15 [a]	1.06 ± 0.05 [a]	1.24 ± 0.09 [a]	0.73 ± 0.11 [b]
TOTAL PHENOLS	449.26 ± 74.39 [a]	332.17 ± 9.24 [b]	414.63 ± 14.35 [ab]	310.06 ± 24.78 [b]	378.19 ± 9.16 [a]	333.22 ± 28.63 [a]	330.32 ± 28.64 [a]	248.42 ± 18.35 [b]

Results are expressed as mean values ± standard deviation of three technical repetitions. Mean values labeled with a different superscript letter, within the same row and same cultivar are statistically different (Tukey's test, $p < 0.05$). In case there were no statistically significant differences the letters were omitted.

Considering the particular secoiridoid compounds in IB oils, most of the concentrations decreased in the oils obtained from stored fruits. Quantitatively the highest reduction with respect to IB-control was determined for the concentration of 3,4-DHPEA-EDA in IB-RT and IB-20 oils. In RO oils from the stored fruits the highest decrease with respect to RO-control oil was detected in the case of 3,4-DHPEA-EDA and oleuropein aglycone (isomer I) after fruit storage at −20 °C and at RT In the oils from both cultivars obtained from fruits stored at +4 °C the profile of secoiridoids was more similar to the control oils than that of the other two treatments. It is probable that the cold storage conditions slowed down the rate of enzymatic and biochemical reactions, which lead to the degradation of these particular phenols, as noticed in RT oils, and at the same time avoided the negative effects caused by freezing, observed in the oils obtained from the fruits stored at −20 °C. These findings are in agreement with the results of Hachicha Hbaieb et al. [9], who found more similar phenolic profiles of the oils obtained from Arbequina fruits stored at 4 °C and the freshly harvested ones, in comparison to the oil obtained from fruits stored at 20 °C, which was explained by the similar endogenous enzyme activity patterns detected in the fruits of the former treatments. Romero et al. [27] characterized the phenolic profile of Spanish olive oils (Cornicabra, Hojiblanca, and Picual cultivars) with "frostbitten olives" sensory defect and found that the concentrations of all the investigated groups of phenols decreased in defective oils, except secoiridoids. The authors [27] explained these differences by considering the action of enzymes that are affected by frost; physical damage of olive fruits by ice crystals formed during freezing leads to cellular destruction, allowing phenolic substrates to mix with polyphenol oxidase (PPO), which degraded them. In this study, lower concentrations of the majority of phenols, even secoiridoids, were found in oils obtained from the fruits frozen at −20 °C in comparison to control oils. This was probably due to the controlled freezing process applied, which did not include freezing and thawing cycles that would correspond to those occurring naturally in the olive orchard.

In both cultivars, a significant increase in the concentrations of simple phenolic compounds, hydroxytyrosol and tyrosol, was found in the oils obtained from the fruits stored at RT and +4 °C compared to the control oils (Table 3). The increase observed was proportional to the storage temperature applied, which was as expected, since it can be explained by increased hydrolysis of complex phenols into simple phenols at higher temperatures [15]. On the other hand, after storage at −20 °C, no significant change in hydroxytyrosol and tyrosol concentrations was found when compared to the control oils. Such an outcome could have possibly been connected to partial inactivation or the lower ability of PPO and peroxidases (PODs) to oxidize biophenolic glucosides at lower storage temperature, as reported earlier [46].

The concentration of total lignans decreased in the case of both monovarietal oils obtained from fruits at −20 °C, which was in agreement with the findings of Masella et al. [26], while in IB+4 oil the increase of total lignans was mainly a consequence of an increase in pinoresinol concentration. Guillaume et al. [42] reported that the concentration of lignans was strongly positively correlated with the intensity of the "frostbitten olives" defect, and that the concentration of acetoxypinoresinol increased after the freezing of olive fruits on the trees, which was not confirmed by the findings of this study in the case of controlled frozen storage.

The concentration of total flavonoids decreased in all the IB treatments, while for RO a decrease was detected only in the case of RO-20 oil. In the oils of both cultivars obtained after storage of fruits at +4 °C the profile of individual flavonoids was more similar to the one observed in the control oils than in the other two treatments. Other authors reported different trends in flavonoids behavior under various storage conditions. Hachicha Hbaieb et al. [9] reported higher flavonoids content in the oils extracted from olives stored at 4 °C than at 20 °C, probably due to the accelerated process of ripening of fruits at the higher temperature. The content of flavonoids in Cornicabra oils obtained from fruits stored at 10 °C and 20 °C did not show a clear trend at the beginning of storage, probably because of their stable structure and high oxidation resistance, while an increase of particular flavonoids was determined after a prolonged storage, probably because of the destruction of the cell structure and the release of bound phenols [37].

The concentration of total phenolic acids only increased after the RT treatment in oils of both cultivars, as a consequence of a sharp increase in p-coumaric acid concentration. Storage at lower temperatures had no influence on the concentration of phenolic acids, which was not in agreement with the results of Masella et al. [26], who reported a decrease in the concentration of p-coumaric acid after 6 months of frozen storage of olive fruits. The discrepancy observed was possibly related to the difference in storage time between the two studies.

4. Conclusions

The results of this study have shown that, when conducted at an appropriate temperature, storage time of olive fruits can be prolonged to seven days without compromising the crucial aspects of olive oil quality. Although prolonged storage at room temperature increased the oil extractability index, this treatment exhibited many serious drawbacks, such as the elevated concentrations of fatty acid ethyl esters and waxes, loss of a certain proportion of valuable phenolic compounds and the occurrence of sensory defects in the obtained oil, most probably due to fermentative processes induced by accelerated post-harvest fruit ripening. Prolonged storage at the freezing temperature of $-20\,^\circ$C also resulted in significant alterations in the composition and quality of the obtained oil, including a decrease in the concentration of phenols and generation of the "frostbitten olives" sensory defect, presumably induced by freezing injuries and modified enzymatic activity in the fruits. The treatment that included prolonged refrigeration of fruits at $+4\,^\circ$C proved to be the most suitable for this purpose, since it preserved the composition and sensory quality most similar to that of the fresh oil of the control treatment, which corresponded to the highest quality category, extra virgin olive oil. The results obtained point to the need to improve olive fruit post-harvest storage technical capabilities and conditions, in order to prevent losses in olive and olive oil quality and value in situations when the harvested amount of fruit exceeds the processing capacity of available mills.

Author Contributions: Conceptualization, K.B.B.; methodology, K.B.B.; formal analysis, K.B.B., I.L., A.N., M.L., M.K.; resources, K.B.B. and I.L.; data curation, K.B.B., M.L., I.L.; writing—original draft preparation, K.B.B.; writing—review and editing, K.B.B., I.L., M.L.; visualization, K.B.B. and M.K.; supervision, K.B.B.; project administration, I.L. and K.B.B.; funding acquisition, I.L. and K.B.B. All authors have read and agreed to the published version of the manuscript.

Funding: This research was funded by Croatian Science Foundation (grant numbers: UIP-2014-09-1194 and DOK-2018-01-4693) and by the European Union from the European Social Fund (grant number: DOK-2018-01-4693).

Conflicts of Interest: The authors declare no conflict of interest. The funders had no role in the design of the study; in the collection, analyses, or interpretation of data; in the writing of the manuscript, or in the decision to publish the results.

References

1. Gómez-Coca, R.B.; Moreda, W.; Pérez-Camino, M.C. Fatty acid alkyl esters presence in olive oil vs. organoleptic assessment. *Food Chem.* **2012**, *135*, 1205–1209. [CrossRef]
2. Vossen, P. Olive oil: History, production and characterization of the world's classic oils. *Hortscience* **2007**, *42*, 1093–1100. [CrossRef]
3. Morales, M.T.; Luna, G.; Aparicio, R. Comparative study of virgin olive oil sensory defects. *Food Chem.* **2005**, *91*, 293–301. [CrossRef]
4. Vichi, S.; Romero, A.; Gallardo-Chacón, J.; Tous, J.; López-Tamames, E.; Buxaderas, S. Influence of olives' storage conditions on the formation of volatile phenols and their role in off-odor formation in the oil. *J. Agric. Food Chem.* **2009**, *57*, 1449–1455. [CrossRef] [PubMed]
5. Perez-Camino, M.C.; Moreda, W.; Mateos, R.; Cert, A. Determination of esters of fatty acids with low molecular weight alcohols in olive oils. *J. Agric. Food Chem.* **2002**, *50*, 4721–4725. [CrossRef] [PubMed]
6. Biedermann, M.; Bongratz, A.; Mariani, C.; Grob, K. Fatty acid methyl and ethyl esters as well as wax esters for evaluating the quality of olive oils. *Eur. Food Res. Technol.* **2008**, *228*, 65–74. [CrossRef]

7. Conte, L.; Bendini, A.; Valli, E.; Lucci, P.; Moret, S.; Maquet, A.; Lacoste, F.; Brereton, P.; García-González, D.L.; Moreda, W.; et al. Olive oil quality and authenticity: A review of current EU legislation, standards, relevant methods of analyses, their drawbacks and recommendations for the future. *Trends Food Sci. Technol.* **2019**, in press. [CrossRef]

8. Regulation, H. European Economic Community. Commission Regulation (EEC) No 2568/91 of 11 July 1991 (and later modifications) on the characteristics of olive oil and olive-residue oil and the relevant methods of analysis. *Off. J. Eur. Communities* **1991**, *L248*, 1–83.

9. Hbaieb, R.H.; Kotti, F.; García-Rodríguez, R.; Gargouri, M.; Sanz, C.; Pérez, A.G. Monitoring endogenous enzymes during olive fruit ripening and storage: Correlation with virgin olive oil phenolic profiles. *Food Chem.* **2014**, *174*, 240–247. [CrossRef]

10. Yousfi, K.; Weiland, C.M.; Garcia, J.M. Responses of fruit physiology and virgin oil quality to cold storage of mechanically harvested 'Arbequina' olives cultivated in hedgerow. *Grasas Aceites* **2013**, *64*, 572–582. [CrossRef]

11. International Olive Council. *Determination of the Content of Waxes, Fatty Acid Methyl Esters and Fatty Acid Ethyl Esters by Capillary Gas Chromatograhy*; COI/T.20/Doc. No 28/Rev.2; International Olive Council: Madrid, Spain, 2017.

12. Boskou, D. *Olive Oil: Chemistry and Technology*, 2nd ed.; AOCS Press: Champaign, IL, USA, 2006.

13. Servilli, M.; Montedoro, G. Contribution of phenolic compounds to virgin olive oil quality. *Eur. J. Lipid Sci. Technol.* **2002**, *104*, 602–613. [CrossRef]

14. Clodoveo, M.L. Malaxation: Influence on virgin olive oil quality. Past, present and future–An overview. *Trends Food Sci. Technol.* **2012**, *25*, 13–23. [CrossRef]

15. Li, X.Q.; Zhu, H.J.; Shoemaker, C.F.; Wang, S.C. The effect of different cold storage conditions on the compositions of extra virgin olive oil. *J. Am. Oil Chem. Soc.* **2014**, *91*, 1559–1570. [CrossRef]

16. Hbaieb, R.H.; Kotti, F.; Cortes-Francisco, N.; Caixach, J.; Gargouri, M.; Vichi, S. Ripening and storage conditions of Chétoui and Arbequina olives: Part II. Effect on olive endogenous enzymes and virgin olive oil secoiridoid profile determined by high resolution mass spectrometry. *Food Chem.* **2016**, *210*, 631–639. [CrossRef] [PubMed]

17. Dag, A.; Boim, S.; Sobotin, Y.; Zipori, I. Effect of mechanically harvested olive storage temperature and duration on oil quality. *Horttechnology* **2012**, *22*, 528–533. [CrossRef]

18. Lukić, I.; Horvat, I.; Godena, S.; Krapac, M.; Lukić, M.; Vrhovsek, U.; Brkić Bubola, K. Towards understanding the varietal typicity of virgin olive oil by correlating sensory and compositional analysis data: A case study. *Food Res. Int.* **2018**, *112*, 78–89. [CrossRef] [PubMed]

19. Brkić Bubola, K.; Koprivnjak, O.; Sladonja, B.; Belobrajić, I. Influence of storage temperature on quality parameters, phenols and volatile compounds of Croatian virgin olive oils. *Grasas Aceites* **2014**, *65*, e034. [CrossRef]

20. García, J.M.; Yousfi, K. The postharvest of mill olives. *Grassas Aceites* **2006**, *57*, 16–24. [CrossRef]

21. Clodoveo, M.L.; Delcuratolo, D.; Gomes, T.; Colelli, G. Effect of different temperatures and storage atmospheres on Coratina olive oil quality. *Food Chem.* **2007**, *102*, 571–576. [CrossRef]

22. Garcia-Vico, L.; Garcia-Rodriguez, R.; Sanz, C.; Perez, A.G. Biochemical aspects of olive freezing-damage: Impact on the phenolic and volatile profiles of virgin olive oil. *LWT-Food Sci. Technol.* **2017**, *86*, 240–246. [CrossRef]

23. Yousfi, K.; Weiland, C.M.; Garcia, J.M. Effect of harvesting system and fruit cold storage on virgin olive oil chemical composition and quality of superintensive cultivated 'arbequina' olives. *J. Agric. Food Chem.* **2012**, *60*, 4743–4750. [CrossRef] [PubMed]

24. Kalua, C.M.; Bedgood, D.R.; Bishop, A.G.; Prenzler, P.D. Changes in virgin olive oil quality during low-temperature fruit storage. *J. Agric. Food Chem.* **2008**, *56*, 2415–2422. [CrossRef] [PubMed]

25. Li, D.M.; Zhu, Z.W.; Sun, D.W. Effects of freezing on cell structure of fresh cellular food materials: A review. *Trends Food Sci. Technol.* **2018**, *75*, 46–55. [CrossRef]

26. Masella, P.; Guerrini, L.; Angeloni, G.; Spadi, A.; Baldi, F.; Parenti, A. Freezing/storing olives, consequences for extra virgin olive oil quality. *Int. J. Refrig.* **2019**, *106*, 24–32. [CrossRef]

27. Romero, I.; Aparicio-Ruiz, R.; Oliver-Pozo, C.; Aparicio, R.; Garcia-Gonzalez, D.L. Characterization of virgin olive oils with two kinds of 'frostbitten olives' sensory defect. *J. Agric. Food Chem.* **2016**, *64*, 5590–5597. [CrossRef]

28. Hbaieb, R.H.; Kotti, F.; Gargouri, M.; Msallem, M.; Vichi, S. Ripening and storage conditions of Chétoui and Arbequina olives: Part. I. Effect on olive oils volatiles profile. *Food Chem.* **2016**, *203*, 548–558. [CrossRef]

29. Poerio, A.; Bendini, A.; Cerretani, L.; Bonoli-Carbognin, M.; Lercker, G. Effect of olive fruit freezing on oxidative stability of virgin olive oil. *Eur. J. Lipid Sci. Technol.* **2008**, *110*, 368–372. [CrossRef]

30. Beltrán, G.; Uceda, M.; Jiménez, A.; Aguilera, M.P. Olive oil extractability index as a parameter for olive cultivar characterisation. *J. Sci. Food. Agric.* **2003**, *83*, 503–506. [CrossRef]

31. Brkić, K.; Radulović, M.; Sladonja, B.; Lukić, I.; Šetić, E. Application of soxtec apparatus for oil content determination in olive fruit. *Riv. Ital. Sostanze Grasse* **2006**, *83*, 115–119.

32. Koprivnjak, O.; Brkić Bubola, K.; Kosić, U. Sodium chloride compared to talc as processing aid has similar impact on volatile compounds but more favorable on ortho-diphenols in virgin olive oil. *Eur. J. Lipid Sci. Technol.* **2016**, *118*, 318–324. [CrossRef]

33. Mínguez-Mosquera, M.I.; Rejano-Navarro, L.; Gandul-Rojas, B.; Sánchez-Gòmez, A.H.; Garrido-Fernandez, J. Colour-pigment correlation in virgin olive oil. *J. Am. Oil Chem. Soc.* **1991**, *68*, 332–336. [CrossRef]

34. International Olive Council. *Sensory Analysis of Olive Oil: Method for the Organoleptic Assessment of Virgin Olive Oil*; COI/T.20/Doc. No 15/Rev.10; International Olive Council: Madrid, Spain, 2018.

35. Jerman Klen, T.; Golc Wondra, A.; Vrhovšek, U.; Mozetič Vodopivec, B. Phenolic profiling of olives and olive oil process-derived matrices using UPLC-DAD-ESI-QTOF-HRMS analysis. *J. Agric. Food Chem.* **2015**, *63*, 3859–3872. [CrossRef] [PubMed]

36. Lukić, I.; Žanetić, M.; Jukić Špika, M.; Lukić, M.; Koprivnjak, O.; Brkić Bubola, K. Complex interactive effects of ripening degree, malaxation duration and temperature on Oblica cv. virgin olive oil phenols, volatiles and sensory quality. *Food Chem.* **2017**, *232*, 610–620. [CrossRef] [PubMed]

37. Inarejos-Garcia, A.M.; Gomez-Rico, A.; Salvador, M.D.; Fregapane, G. Effect of preprocessing olive storage conditions on virgin olive oil quality and composition. *J. Agric. Food Chem.* **2010**, *58*, 4858–4865. [CrossRef] [PubMed]

38. Garcia, J.M.; Gutierrez, F.; Barrera, M.J.; Albi, M.A. Storage of mill olives on an industrial scale. *J. Agric. Food Chem.* **1996**, *44*, 590–593. [CrossRef]

39. Garcia, J.M.; Gutiérrez, F.; Castellano, J.M.; Perdiguero, S.; Morilla, A.; Albi, M.A. Influence of storage temperature on fruit ripening and olive oil quality. *J. Agric. Food Chem.* **1996**, *44*, 264–267. [CrossRef]

40. Morello, J.R.; Motilva, M.J.; Ramo, T.; Romero, M.P. Effect of freeze injuries in olive fruit on virgin olive oil composition. *Food Chem.* **2003**, *81*, 547–553. [CrossRef]

41. Kiritsakis, A.; Nanos, G.D.; Polymenopoulos, Z.; Thomai, T.; Sfakiotakis, E.M. Effect of fruit storage conditions on olive oil quality. *J. Am. Oil Chem. Soc.* **1998**, *75*, 721–724. [CrossRef]

42. Guillaume, C.; Ravetti, L.; Gwyn, S. Characterisation of phenolic compounds in oils produced from frosted olives. *J. Am. Oil Chem. Soc.* **2010**, *87*, 247–254. [CrossRef]

43. Romero, I.; Garcia-Gonzalez, D.L.; Aparicio-Ruiz, R.; Morales, M.T. Study of volatile compounds of virgin olive oils with 'frostbitten olives' sensory defect. *J. Agric. Food Chem.* **2017**, *65*, 4314–4320. [CrossRef]

44. Jabeur, H.; Zribi, A.; Abdelhedi, R.; Bouaziz, M. Effect of olive storage conditions on Chemlali olive oil quality and the effective role of fatty acids alkyl esters in checking olive oils authenticity. *Food Chem.* **2015**, *169*, 289–296. [CrossRef] [PubMed]

45. Bendini, A.; Cerretani, L.; Carrasco-Pancorbo, A.; Gómez-Caravaca, A.M.; Segura-Carretero, A.; Fernández-Gutiérrez, A.; Lercker, G. Phenolic molecules in virgin olive oils: A survey of their sensory properties, health effects, antioxidant activity and analytical methods. An overview of the last decade. *Molecules* **2007**, *12*, 1679–1719. [CrossRef] [PubMed]

46. Taticchi, A.; Esposto, S.; Veneziani, G.; Urbani, S.; Selvaggini, R.; Servili, M. The influence of the malaxation temperature on the activity of polyphenoloxidase and peroxidase and on the phenolic composition of virgin olive oil. *Food Chem.* **2013**, *136*, 975–983. [CrossRef] [PubMed]

Article

An Artificial Intelligence Approach for Italian EVOO Origin Traceability through an Open Source IoT Spectrometer

Simona Violino [1], Luciano Ortenzi [1], Francesca Antonucci [1], Federico Pallottino [1,*], Cinzia Benincasa [2], Simone Figorilli [1] and Corrado Costa [1]

[1] Consiglio per la Ricerca in Agricoltura e l'analisi dell'economia Agraria (CREA)—Centro di Ricerca Ingegneria e Trasformazioni Agroalimentari, Via della Pascolare 16, 00015 Monterotondo (Rome), Italy; simonaviolino@hotmail.com (S.V.); luciano.ortenzi@crea.gov.it (L.O.); francesca.antonucci@crea.gov.it (F.A.); simone.figorilli@crea.gov.it (S.F.); corrado.costa@crea.gov.it (C.C.)

[2] Consiglio per la Ricerca in Agricoltura e l'analisi dell'economia Agraria (CREA)—Centro di Ricerca Olivicoltura, Frutticoltura e Agrumicoltura, Contrada Li Rocchi Vermicelli 83, 87036 Rende (CS), Italy; cinzia.benincasa@crea.gov.it

* Correspondence: federico.pallottino@crea.gov.it; Tel.: +39-06-906-75-268

Received: 8 June 2020; Accepted: 17 June 2020; Published: 25 June 2020

Abstract: Extra virgin olive oil (EVOO) represents a crucial ingredient of the Mediterranean diet. Being a first-choice product, consumers should be guaranteed its quality and geographical origin, justifying the high purchasing cost. For this reason, it is important to have new reliable tools able to classify products according to their geographical origin. The aim of this work was to demonstrate the efficiency of an open source visible and near infra-red (VIS-NIR) spectrophotometer, relying on a specific app, in assessing olive oil geographical origin. Thus, 67 Italian and 25 foreign EVOO samples were analyzed and their spectral data were processed through an artificial intelligence algorithm. The multivariate analysis of variance (MANOVA) results reported significant differences ($p < 0.001$) between the Italian and foreign EVOO VIS-NIR matrices. The artificial neural network (ANN) model with an external test showed a correct classification percentage equal to 94.6%. Both the MANOVA and ANN tested methods showed the most important spectral wavelengths ranges for origin determination to be 308–373 nm and 594–605 nm. These are related to the absorption of phenolic components, carotenoids, chlorophylls, and anthocyanins. The proposed tool allows the assessment of EVOO samples' origin and thus could help to preserve the "Made in Italy" from fraud and sophistication related to its commerce.

Keywords: VIS-NIR; ANN; made in Italy; minor components; pigments; antioxidants; non-destructive techniques; ready-to-use; spectral signature; artificial intelligence AI

---

1. Introduction

Extra virgin olive oil (EVOO) represents one of the most important ingredients of the Mediterranean diet, being used by most of the countries within the Mediterranean basin, owing to its excellent qualities and sensory properties ascribable to the fruits of olive trees (*Olea europaea* L.) [1]. The qualitative characteristics and the taste of EVOO are largely influenced by the olive plant varieties, the geographical origin, and the agronomic and production techniques employed as well. [2]. Recently, the consumption of EVOO has increased worldwide, even outside the Mediterranean and European countries (for example, India, Russia, China, and Australia). This trend demonstrates an increasing interest of both the producers and the consumers on the quality of food and calls for proper geographical identification and traceability of EVOOs [3].

The price of EVOO is on average 4–5 times higher than other vegetable oils. This is due to the higher production costs and to its higher nutritional and organoleptic properties. Therefore, the higher cost should in principle help to ensure high quality standards. On the other hand, the consumer is increasingly oriented towards the purchase of genuine food products with certified geographical origin [4]. In order to preserve the EVOO origin, the European Commission has established two types of certification relative to geographical origin and identification, namely the protected designation of origin (PDO) and the protected geographical indication (PGI) [4]. The definition of PGI refers to agricultural products and foodstuffs for which at least one stage of the production process must be carried out within a defined geographical area. For PDO, on the other hand, the entire production cycle must take place in a specific area. PGI labelling, therefore, focuses on quality and specific characteristics related to geographical origin [5]. As reported by the production regulations, in order to obtain the PDO certification, several conditions must be met such as a specific percentage of olive cultivars employed, well-defined cultivation practices, limited geographical areas of production, and specific characteristics regarding chemical and sensory properties of the final product. However, at the moment, to the best of our knowledge, there are no analytical parameters allowing a post hoc test on the actual geographical origin of PDO EVOOs. As a consequence, chemical and physical analyses are currently of limited use in the EVOO geographical certification [6].

Italy is one of the most important countries in the world in terms of olive oil supply and demand. Moreover, while the country boasts a large number of designations of origin labels, in total there are indeed forty-two PDOs of EVOO, the geographical indications, PGIs, exist for a limited number of three. The most active regional Italian realities are mainly located in the southern part of the peninsula, namely the region of Sicily with six PDOs and one PGI, Puglia with five PDOs, and Campania and Calabria with three PDOs each; however, there are also two important Italian central regions, such as Lazio and Tuscany, with four PDOs and one PGI, and five PDOs and one PGI, respectively [7].

Unfortunately, nowadays, one of the most common counterfeits to the detriment of the producer is the falsification of the EVOO geographical origin. Despite the great commitment and work of the authorities, counterfeiting is highly relevant for both the internal market and the global one, with difficulties in identifying and adopting convenient and reliable solutions. Indeed, due to lack of incentives, and often complex management to comply with, the technologies promoting product traceability are often difficult to implement.

Given the presence of consumers not aware of the fakes, an increasing number of low-quality olive oils often end up on the table, being not easily identifiable. The problems do not only concern consumers but also producers who operate correctly who are not economically damaged by the irregular practices of other companies [7]. Nevertheless, this type of counterfeiting causes enormous damages to the "Made in Italy" products' image and to the economy of the country.

Fake EVOO bottles often report on the label incorrect information about the product or even refer to a totally different oil.

The "Made in Italy" products are represented by a set of values enabling the consumer to distinguish them from the foreign ones. However, sometimes their advertisement is used to lead the consumer to pay an even higher price for fake qualities relative to a forged product [8]. Thus, the consumer must also be aware of the differences between fake and authentic products, being guided through tools that allow them to distinguish "what is" from "what appears to be" [4].

However, within a globalized market, the fight against these counterfeits cannot be solely based on the enhancement of consumers' awareness on the peculiarities and qualities that distinguish the "Made in Italy" from other products. Useful tools to contend with counterfeiting are those ensuring traceability [9], namely, the "possibility of reconstructing and following the path of a food in all phases of production, transformation and distribution" [10]. Therefore, traceability systems (technological and informative) are needed to strengthen and update a reliable information flow along the whole supply chain, simplifying consumers' access to information. An infotracing traceability system can integrate information related to product quality with that regarding its traceability (physical and documentary), taking advantage of an online information system [9]. As reported by Violino et al. [11], EVOO traceability is not only important to define olive oil origin, but it is fundamental for the protection against fraud. In addition, innovative tools (e.g., radio frequency identification (RFID), near field communication (NFC), and QR code technologies in combination with blockchain systems) can be commercially implemented to verify the processes and could aid in controlling the quality of virgin olive oil. An example is represented by the use of the blockchain system (a distributed database of records in the form of encrypted blocks), where the developed online information (transactions) can be protected to proof against eventual alteration and fraud [12].

Nowadays, the traceability of food products has become a priority for both consumers, who are increasingly careful to buy healthy higher quality food, and producers. Indeed, traceability can guarantee the quality of raw materials, product certification, allow a rapid identification of problematic product lots, and permit the implementation of control systems to prevent fraud. Finally, food traceability is crucial to enhance transparency for a safer internationalization of the EVOO market, with consequent fair growth of the sector [13].

In the last decade, several analytical techniques have been developed to help the identification of olive oil [14,15], and about 200 compounds, out of hundreds, have been proved to be useful as compositional markers for traceability purposes of EVOO [16]. Compositional markers include both major and minor components. State of the art EVOO traceability approaches for geographical origin assessment are represented by major components determination (e.g., triacylglycerols, triglycerides, and fatty acids), stable isotopic ratio (e.g., $^{13}C/^{12}C$ in combination with $^{18}O/^{16}O$), and multi-element characterization through the application of different multivariate statistical techniques [3]. Those commonly used for data analysis are cluster analysis, multidimensional scaling, artificial neural networks (ANNs), and partial least squares discriminant analysis (PLSDA).

The multi-element analysis carried out by Benincasa et al. [17] allowed for a correct classification of all the organic virgin olive oils under investigation collected from different Italian regions; however, as often visible in similar studies, the method showed a high, but not excellent, percentage of correct classifications. Another example is given by the stable isotope analysis made by Portarena et al. [18], reporting an *r* ranging from 0.76 to 0.80 in distinguishing the compositions of Italian monovarietal olive oils.

Numerous analytical techniques have focused on targeted approaches for the identification and quantification of pre-defined compounds, or classes of compounds. These include gas and liquid chromatography (GC and HPLC) coupled with mass spectrometry (MS) [19,20], nuclear magnetic resonance (NMR) spectroscopy [21], infrared spectroscopy [22], fluorescence [23], inductively coupled plasma mass spectrometry (ICP-MS) [17], and DNA-based methods [24]. Conversely, limited literature is available about the assessment of olive oil adulteration using non-targeted classification approaches, focusing on the detection of all compounds in a sample without a priori knowledge of chemical entities comparable with the reference of the pure sample fingerprint profile [25,26].

Recently, fundamental research has focused on the development of non-destructive techniques to reduce the use of solvents and reagents. This is done taking into account an international context of convergence towards higher environmental sustainability and an increased human health consciousness [27]. Among various non-destructive solutions aiming to fulfill these needs, near infrared spectroscopy (NIRS) has made major achievements. NIRS, paired with chemometric techniques, were satisfactorily used for olive oil authentication and screening [28–33]. Generally, using both software and hardware, open source infostructure solutions potentially result in significant cost reduction, making the scientific tools available for a wider audience [34]. Following this path, results such as the prediction of qualitative parameters, the evaluation of indices of different fruit and vegetable products [35], the authentication of olive oil according to the variety and geographical origin [36], and the detection of adulteration through acidic composition [37] were achieved using visible/near-infrared (VIS-NIR) spectroscopy.

The aim of this work was to assess the actual geographical origin of EVOOs labeled on the market as Italian and test the potential efficiency of an open source VIS-NIR device for traceability purposes. Indeed, the device could produce results for olive oil authentication (according to its variety and origin) and for the detection of fraud in a fraction of time and potentially on a much higher sampling number with respect to conventional analytical methods. In detail, the study pursued the goal of analyzing 92 Italian and foreign EVOO samples produced in 2018 and 2019. The samples were purchased from large commercial retailers and directly from olive mills (to ensure the true origin of the product). The spectral data were analyzed with an artificial intelligence model based on neural networks.

2. Materials and Methods

2.1. EVOO Samples

The study analyzed a total of 92 samples of Italian and foreign extra virgin olive oil (EVOO) owing to different cultivars, monovarietal (65) and blend (27), produced in two harvest years (2018 and 2019) (Figure 1).

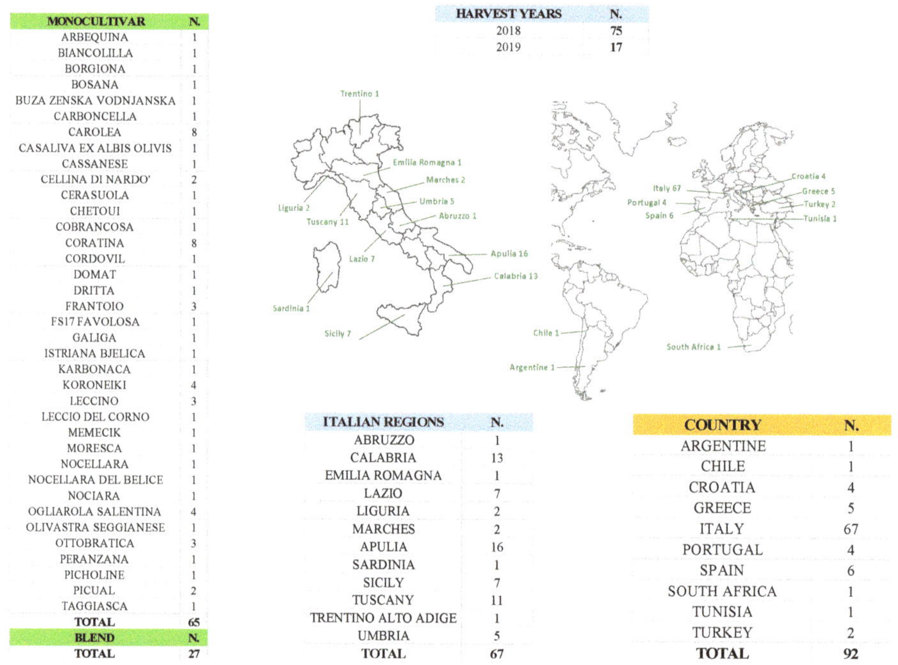

MONOCULTIVAR	N.
ARBEQUINA	1
BIANCOLILLA	1
BORGIONA	1
BOSANA	1
BUZA ZENSKA VODNJANSKA	1
CARBONCELLA	1
CAROLEA	8
CASALIVA EX ALBIS OLIVIS	1
CASSANESE	1
CELLINA DI NARDO'	2
CERASUOLA	1
CHETOUI	1
COBRANCOSA	1
CORATINA	8
CORDOVIL	1
DOMAT	1
DRITTA	1
FRANTOIO	3
FS17 FAVOLOSA	1
GALIGA	1
ISTRIANA BJELICA	1
KARBONACA	1
KORONEIKI	4
LECCINO	3
LECCIO DEL CORNO	1
MEMECIK	1
MORESCA	1
NOCELLARA	1
NOCELLARA DEL BELICE	1
NOCIARA	1
OGLIAROLA SALENTINA	4
OLIVASTRA SEGGIANESE	1
OTTOBRATICA	3
PERANZANA	1
PICHOLINE	1
PICUAL	2
TAGGIASCA	1
TOTAL	**65**
BLEND	**N.**
TOTAL	**27**

HARVEST YEARS	N.
2018	75
2019	17

ITALIAN REGIONS	N.
ABRUZZO	1
CALABRIA	13
EMILIA ROMAGNA	1
LAZIO	7
LIGURIA	2
MARCHES	2
APULIA	16
SARDINIA	1
SICILY	7
TUSCANY	11
TRENTINO ALTO ADIGE	1
UMBRIA	5
TOTAL	**67**

COUNTRY	N.
ARGENTINE	1
CHILE	1
CROATIA	4
GREECE	5
ITALY	67
PORTUGAL	4
SPAIN	6
SOUTH AFRICA	1
TUNISIA	1
TURKEY	2
TOTAL	**92**

Figure 1. Monocultivar and blend extra virgin olive oil (EVOO) samples.

The tested samples were bought from large retailers and directly from mills. Some samples were acquired specifically from the mills of the areas of Apulia, Calabria, and Sicily to ensure their origin. Other samples were sent, on a voluntary base, directly by the producers willing to participate in the research.

2.2. The Open Source IoT Spectrometer

The analyzed samples were stored and kept during the analyses at a controlled temperature of 16 °C. The samples, owing to the 2018 harvest campaign, were analyzed between March and May

2019 while those produced in the 2019 harvest campaign were analyzed between February and March 2020. The samples were scanned with a VIS-NIR spectrometer measuring and acquiring the spectral reflectance signatures for the EVOO samples for consequent qualitative evaluation. From each oil container (bottle or can) of the same sample, 12 spectral readings were acquired and afterwards averaged. The device used was the ultra-compact VIS-NIR spectrophotometer (Figure 2) Lumini C (Myspectral Ltd., Cambridge, MA, USA), able to measure spectral reflectance or absorbance. The device is small, light, low-cost, and open source. The spectral ranges covered 340–890 nm with an optical resolution equal to 8 nm and wavelength accuracy equal to 0.5 nm. The spectrophotometer is powered through a USB cable and stores data on connected cabled devices or on an internal micro SD card using a dedicated slot. For appropriate acquisition of the spectral signature, in relation to the sample reflectance characteristics, the acquisition can be set at different integration times. The system is equipped with its own internal illumination system.

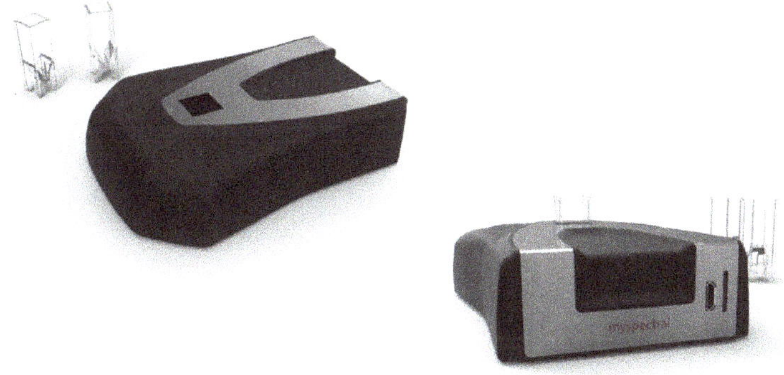

Figure 2. VIS-NIR ultra-compact spectrophotometer Lumini C Myspectral using standard cuvette holder for absorbance spectrophotometry.

A specific app was developed to manage and simplify the acquisition procedures. The software provided with the spectrophotometer, as commonly happens with open source technologies, was quite poor in terms of features and did not originally provide an appropriate historicization system for multiple acquisitions. For this reason, an app was developed and implemented. A screenshot of the app is reported in Figure 3.

The app was engineered considering two kinds of functions. The first (upper side of Figure 3) enables the configuration parameters of the instrument, such as the IP address, to connect the tablet to the device, the type of tool (in this case is Lumini C), the exposure time expressed in milliseconds (ms), and the sample's name to be archived. The second (lower side of Figure 3), graphically represents the acquired spectrum for each scan. When a new sample name is entered, the graphic area is reset, ready to display the new spectra. This helps in case of incomplete or bad acquisition since it avoided losing samples' values during the acquisition campaign. The app was developed using the Android environment and it is based on a client-server paradigm; on the client side there is the app, and on the server side there is the database for real-time storage of the spectrum and the node.js server to which the Lumini C is connected (Figure 4). The app software implements control mechanisms for the data stored on the database; these are essential since the data stored originally onboard within a microSD are now stored to a remote database. Through this mechanism, the data loss is minimized. In case of communication problems among the devices, the app notifies the problem and does not display the spectrum just acquired, allowing for a new scanning process.

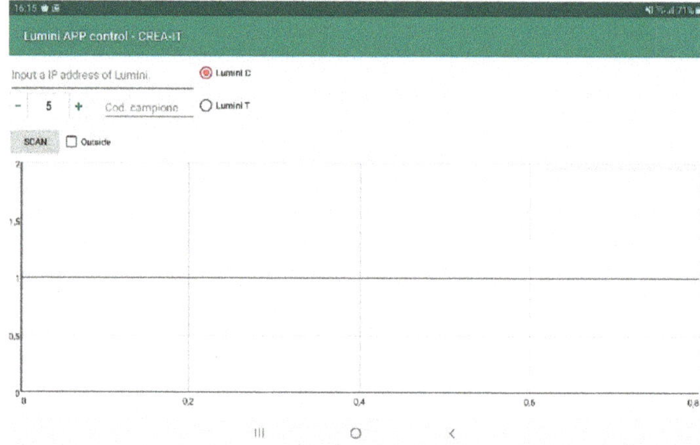

Figure 3. Screenshot of the Lumini app control CREA-IT for spectrophotometric acquisitions of EVOO samples.

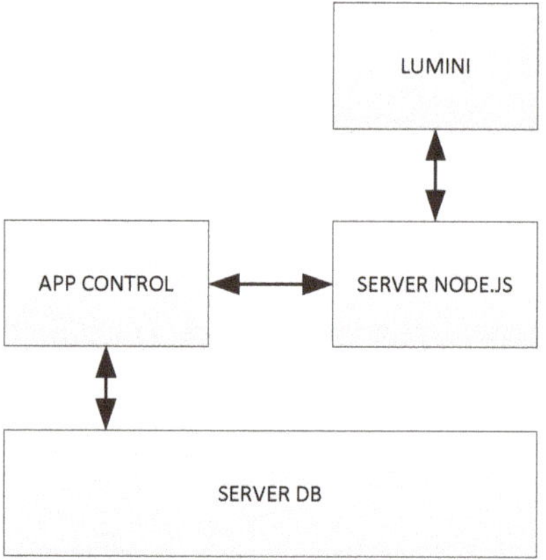

Figure 4. Block diagram of the Lumini C acquisition system via Android app.

2.3. Statistical Analysis

The multivariate matrix of Italian and foreign EVOO samples was analyzed with a 50–50 multivariate analysis of variance (MANOVA) procedure [38], a generalized multivariate Anova method based on principal component analysis (PCA) standardized data. The MANOVA was conducted in order to highlight significant differences between Italian and foreign VIS-NIR matrices. Adjusted p-values were conducted on a rotation testing based on 99,999 simulated datasets. The contribution of the variables was extracted for each rotation test [39].

An artificial intelligence approach was then applied in order to evaluate the possibility to classify Italian EVOOs and distinguish them from the foreign ones on the base of the 288 spectral transmittance values acquired through the VIS-NIR device. To do this, a multilayer feed forward artificial neural network (MLFN) was designed using a single hidden layer architecture with sigmoid hidden and SoftMax output neurons. The ANN was trained with the Bayesian regularization back propagation algorithm [40,41], as implemented in the deep learning MATLAB (The MathWorks, Inc., MA, USA) toolbox. The dataset was partitioned using 60 percent of the samples (55) as a training set and the rest as a test set (37). The test set was used to validate the model. This partitioning (equal for each soil group) was optimally chosen with the Euclidean distances calculated by the algorithm reported by Kennard and Stone [42], selecting parameters without a priori knowledge of a regression model. The cost function was minimized using the root mean squared (RMS) normalized error performance function with a 10^{-8} threshold on the gradient. In order to extract the most informative spectral transmittance values among the 288 acquired, in distinguishing Italian EVOO from foreign ones, it also conducted an analysis to study the feature importance. The hidden layer matrix (10 nodes × 288 variables) was a posteriori analyzed considering its elementwise absolute value. From the matrix was extracted the maximum value for each variable (e.g., column) obtaining a 1 × 288 row vector. The top 40 most significant spectral frequencies were chosen. The larger the value, the more relevant was the contribution to the ANN model. The model was developed using the MATLAB 9.7 R2019b Deep Learning Toolbox.

3. Results and Discussion

3.1. Artificial Intelligence Modeling Based on VIS-NIR Spectra

The MANOVA (50–50 MANOVA procedure) reported significant differences ($p < 0.001$) between the two Italian and foreign EVOO VIS-NIR matrices. The results of the analysis are reported in Table 1.

Table 1. MANOVA results based on Italian and foreign EVOO samples.

Source	DF	exVarSS	nPC	nBu	exVarPC	exVarBU	*p*-Value
Italian vs. Foreign	1	0.04276	2	42	0.832	1	0.005056
Error	90	0.95724					

DF, degrees of freedom; exVarSS, explained variances based on sums of squares; nPC, number of principal components used for testing; nBu, number of principal components used as buffer components; exVarPC, variance explained by nPC components; exVarBU, variance explained by (nPC+nBU) components; *p*-value, the result from 50–50 MANOVA testing.

The ANN trained had a hidden layer size of 10 nodes and the algorithm converged after 976 iterations. Table 2 reports the characteristics and principal results of the ANN model used to predict Italian vs. foreign EVOO on the base of 288 VIS-NIR spectral transmittance data. All the 55 EVOOs in the training set were correctly classified. In testing, only five out of 37 samples were misclassified. These five samples consisted of two Italian commercial monocultivars (Coratina from Apulia and Taggiasca from Liguria) and three foreign blends from Greece, Argentina, and Croatia. Overall, 87 out of 92 samples (94.6%) were correctly classified.

Table 2. Characteristics and principal results of the multilayer feed forward artificial neural network (MLFN) model (training and internal test) in predicting the classification of Italian vs. foreign EVOO: number of cases, training time, number of trials, and percentage of bad predictions.

Training (60%)	
Number of Cases	55
Number of hidden layers	1
Number of nodes	10
Training time	1:26:02
Number of trials	976
% bad predictions	0.0
Testing (40%)	
Number of cases	37
% bad predictions (N)	13.51 (5)

The confusion matrix of the test set is reported in Table 3.

Table 3. Confusion matrix of the test set of the MLFN model used in predicting the classification of Italian vs. foreign EVOO. The correctly classified samples are reported on the main diagonal of the matrix.

	Italian	**Foreign**	**Total**
Italian	25	2	24
Foreign	3	7	8

Overall, VIS-NIR spectroscopy analyses showed significant differences between Italian and foreign samples. From the results obtained through the ANN analysis, only five samples out of 37 were misclassified, e.g., two Italian commercial monocultivars (Coratina from Apulia and Taggiasca from Liguria) and three foreign blends (from Greece, Argentina, and Croatia). Probably, the two Italian samples were misclassified because of their uncertain geographical origin, considering that they are commercial oils. All the samples bought directly from the mills (noncommercial) were correctly classified. The off diagonal elements of the test confusion matrix (Table 3) are reported in Table 4.

Table 4. Off diagonal elements of the test confusion matrix reported in Table 3.

Origin	**Cultivar**	**Commercial**
Italy	Coratina	Yes
Italy	Taggiasca	Yes
Greece	Koroneiki	Yes
Argentina	Coratina	Yes
Croatia	Karbonaka	Yes

Generally, machine learning relies on the amount of data for good modeling, where more data correspond to a modeling approach with increased robustness and performance. For this reason, even if the overall accuracy of the model is almost 90% and the convergence threshold of 10-8 on the RMS error gradient is very strict, the small size of the dataset (made of 92 samples) is not enough to validate the model. On the other hand, the high accuracy obtained despite the small dataset returns the reliability of the correlation observed [43].

The present work considered 67 Italian EVOOs and 25 foreign ones (two harvesting years: 2018 and 2019). However, it must be considered that other work using different methods to authenticate EVOO geographical origin were developed using a number of samples comparable and sometimes lower than that presented in this work. As reported by Bucci et al. [44], the data set for the statistical analysis was constructed on the results of the chemical analyses performed on 153 EVOOs (years of harvesting: 1997–1999), but finally only the samples produced in 1999 (53 oils) were analyzed in the laboratory. In the work conducted by Portarena et al. [45], they analyzed the isotopic composition and carotenoid content of 38 EVOOs from seven regions along the Italian coast using isotope ratio mass spectrometry (IRMS) and resonance Raman spectroscopy (RRS). The correlation between color and pigment content is well known in the literature [46]: the crushing of very green olives produces a typical green colored oil due to the high content in chlorophyll; if olives are more mature, carotenoids will prevail, determining a yellow-gold colored oil. Additionally, as the maturation progresses, the content and profile of phenolic compounds will also be affected: crushing green olives will result in an oil characterized by a higher content of phenolic acids, phenolic alcohols, oleuropein, and secoiridoids, whereas oils produced with dark brown olives will have a high content of anthocyanins, water-soluble plant pigments that take on different colors: red, blue, or violet [47,48].

3.2. Feature Importance

Observing the top 40 MANOVA rotation test's most important variables (e.g., spectral lengths), the most informative ones ranged within the following frequencies: 308–373 nm, 594–612 nm, and 617–641 nm. The average VIS-NIR spectral data of foreign and Italian EVOOs are reported in Figure 5 together with the higher importance spectral values extracted with the aforementioned MANOVA rotation test.

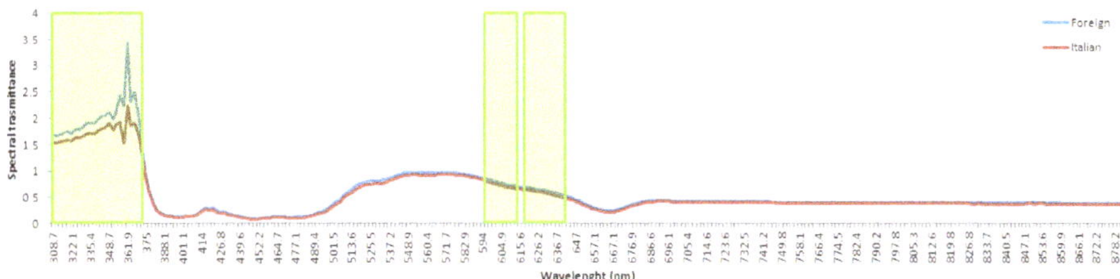

Figure 5. Mean VIS-NIR spectral data: Italian (red line), foreign (blue line). Higher importance spectral values extracted with the MANOVA rotation test are evidenced with green rectangles.

Consequently, the 40 most important features extracted through the ANN procedure (e.g., spectral lengths), ranged within the following frequencies: 308–378 nm, 415–422 nm, 474–507 nm, 564–570 nm, and 596–605 nm. The average VIS-NIR spectral data of both Italian and foreign samples, together with the higher importance spectral values in terms of ANN feature importance, are reported in Figure 6.

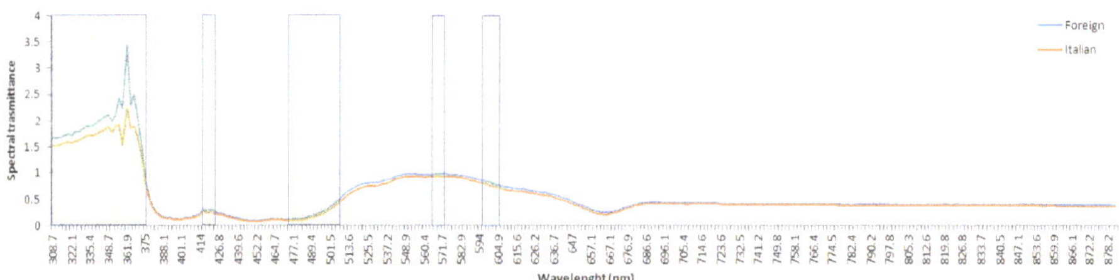

Figure 6. Mean VIS-NIR spectral data: Italian (red line), foreign (blue line). Higher importance spectral values in terms of artificial neural network (ANN) feature importance are evidenced with green rectangles.

The two feature importance approaches, MANOVA and ANN, evidenced common ranges of higher importance, which were: 308–373 nm and 594–605 nm. These spectral bands represent portions of the visible spectral range. The color of an oil is, therefore, due to the combination and proportion of its pigments [49]. These molecules do not depend only on the characteristics of the fruits (*Olea europaea* L.), the extraction processes used to produce the oil, and the conservation conditions [50] but, also, on weather and pedo-climate conditions [51]. Therefore, the relationship between the stage of ripeness and pigment content in EVOO could be, indeed, very important for further authentication studies [52].

The molecular structure of chlorophylls and, in particular, the planar structure of the tetrapyrrolic macrocycle coordinated by a magnesium ion, Mg^{++}, is responsible for the absorption of visible light in the green region. Chlorophyll a gives a greenish-blue coloration, while chlorophyll b determines a yellowish-green color. The sensitivity of chlorophylls to extreme temperature and pH allows the formation of several distinct derivatives such as pheophytins, chlorophyllides, and pheophorbides. During the olive oil extraction process, the release of acids may cause pheophytinization reactions in the chlorophyll fraction, increasing the oils' pheophytin content. The conversion of chlorophylls to Mg^{2+} free derivatives, such as pheophytins, where the Mg^{++} ion is replaced by two H^+ ions, causes oil color changes over time [53–55]. Pheophytin a is present in greater quantities than pheophytin b.

If olive oil is not well preserved, pheophytins can transform further, degrading to pyro pheophytin [55]. These latter can be considered an index of an aging oil. In addition to chlorophyll derivatives, pigments in extra virgin olive oil include carotenoids, the majority of which are lutein and carotene. Carotenoids are isoprenoid compounds with a hydrocarbon structure with various double bonds, C–C, which are responsible for their interesting properties as antioxidants [56]. Carotenoids can be further divided into carotenes (which contain only carbon and hydrogen atoms) and xanthophylls (which also contain oxygen atoms).

The spectra of olive oils analyzed in this work agree with those reported in the literature [57–61].

The peaks occurring in the range between 308 and 380 nm are mostly due to phenolic components [62]. In detail, we found the peak at around 350 nm, the absorption zone of flavones, present in the EVOO absorbance spectrum useful to distinguish Italian EVOOs from foreign ones. Flavonoids are plant secondary metabolites with different phenolic structures. These compounds are mostly used to generate pigments, which play an important role in the colors of plants producing yellow or red/blue pigmentation. Flavonoids such as apigenin, apigenin-7-*O*-glucoside, luteolin, luteolin-7-*O*-glucoside, luteolin-4-*O*-glucoside, diosmetin, quercetin, and quercetin-3-rutinoside are present in olive oils and contribute to the health benefits of consumers. The antioxidant and cellular damage repairing properties that make them useful for preventing cancer, cardiovascular disease, and degenerative diseases in general have been widely studied [63]. The main factors that contribute to their increase in the oil is the maturity index of the fruits and the degree of grinding and the malaxation conditions of the paste during the extraction processes of the oil [64].

The peaks occurring in the range between 415 and 422 nm are due to the compounds absorbing dark blue colored light, mainly carotenoids, as well as pheophytin a, pheophorbide a, and pyro pheophytin a [59], and are characterized by a yellow color.

The peaks occurring in the range between 474 and 507 nm are due to the compounds absorbing green/yellow colored light, and correspond to carotenoids, such as astaxanthin and canthaxanthin. In any case, the major carotenoids in olive oil are β-carotene and lutein, both of them providing several health benefits. Lutein exhibits antioxidant and anti-inflammatory activity protecting against DNA damage [65].

Moreover, the peaks occurring in the range between 564 and 570 nm and between 594 and 605 nm are due to the compounds absorbing orange colored light, characterized by purple/violet and green/blue colors, respectively, and corresponding to chlorophylls and anthocyanins.

4. Conclusions

Spectroscopic techniques paired to chemometric analyses are widely used to authenticate and differentiate edible oils. Most spectroscopic methods tend to focus on the major compounds of the saponifiable fraction of an oil, and only a few have been concentrating on the contents of minor compounds, such as pigments and antioxidants. The European community has not yet accepted many of the scientific community's indications concerning minor compounds, which, by law, are not taken into consideration for the definition of EVOOs' authenticity. However, many of the minor compounds are present in significant amounts only in EVOOs, and their quantification could greatly help the oil industry. Although further analysis will be needed to expand the case studies on olive oils, this work provides a clear indication of how pigment and antioxidant contents are crucial for the authentication and definition of the quality parameters of an EVOO. In detail, we found that the peak at about 360 nm and the broad band around 550 nm present in the EVOO absorbance spectrum can be used to distinguish Italian EVOO from foreign ones. As opposite to expensive and time-consuming chromatographic methods, procedures relying on (open source) spectroscopic instruments are cheap (less than 1000 €) and do not require sample preprocessing. Moreover, being fast, these techniques can be used to assess a huge collection of samples within a reasonable time. The quantitative analysis of pigments can take place directly at production sites and stores, through portable tools that are easy to use, even by non-expert staff. The trained ANN used to classify the samples according to their optical spectra can be easily implemented on an app for immediate classification. The development of simple and reliable methods that can verify the authenticity and guarantee the quality of agri-food products is crucial. Encouragingly, this type of analysis would be very beneficial for the producers themselves as well as consumers. Indeed, these techniques can score comparable precision with respect to the more expensive and time-consuming traditional ones. Moreover, since their application cost relies entirely on the instrumental budget, and not on reagent or other expensive consumable materials, they can be applied to a high number of samples and thus, in case of supposed fraud, can be used as pre-screening tools leading to time and economic optimization.

Author Contributions: Conceptualization, F.P. and C.C.; data curation, C.C., S.V., and L.O.; formal analysis, C.C., S.V., and L.O.; funding acquisition, F.P. and C.C.; investigation, S.V., F.P., L.O., C.B., and C.C.; methodology, S.F., S.V., L.O., and C.C.; project administration, C.C.; software, S.F. and L.O.; supervision, F.P. and C.C.; validation, L.O. and C.C.; visualization, L.O., C.B., and C.C.; writing—original draft, S.V., F.A., C.B., F.P., and C.C.; writing—review and editing, F.P. and S.V. All authors have read and agreed to the published version of the manuscript.

Funding: This research was funded by the Italian Ministry of Agriculture (MiPAAF), grant number D.M. n.12479 of project INFOLIVA.

Acknowledgments: The authors would like to acknowledge all the farms which sent the EVOO to be analyzed and Sara Yazji for illuminating discussions on the VIS-NIR absorbance spectra.

Conflicts of Interest: The authors declare no conflict of interest.

References

1. Benito, M.; Oria, R.; Sánchez-Gimeno, A.C. Characterization of the olive oil from three potentially interesting varieties from Aragon (Spain). *Food Sci. Technol. Int.* **2010**, *16*, 523–530. [CrossRef] [PubMed]
2. Bevilacqua, M.; Bucci, R.; Magrì, A.D.; Magrì, A.L.; Marini, F. Tracing the origin of extra virgin olive oils by infrared spectroscopy and chemometrics: A case study. *Anal. Chim. Acta* **2012**, *717*, 39–51. [CrossRef] [PubMed]
3. Chiavaro, E.; Cerretani, L.; Di Matteo, A.; Barnaba, C.; Bendini, A.; Iacumin, P. Application of a multidisciplinary approach for the evaluation of traceability of extra virgin olive oil. *Eur. J. Lipid Sci. Technol.* **2011**, *113*, 1509–1519. [CrossRef]
4. Violino, S.; Pallottino, F.; Sperandio, G.; Figorilli, S.; Antonucci, F.; Ioannoni, V.; Fappiano, D.; Costa, C. Are the innovative electronic labels for extra virgin olive oil sustainable, traceable, and accepted by consumers? *Foods* **2019**, *8*, 529. [CrossRef]
5. Girelli, C.R.; Del Coco, L.; Zelasco, S.; Salimonti, A.; Conforti, F.L.; Biagianti, A.; Barbini, D.; Fanizzi, F.P. Traceability of "Tuscan PGI" extra virgin olive oils by 1H NMR metabolic profiles collection and analysis. *Metabolites* **2018**, *8*, 60. [CrossRef]
6. Cosio, M.S.; Ballabio, D.; Benedetti, S.; Gigliotti, C. Geographical origin and authentication of extra virgin olive oils by an electronic nose in combination with artificial neural networks. *Anal. Chim. Acta* **2006**, *567*, 202–210. [CrossRef]
7. Ferronato, M. Il Mercato dell'Olio di Oliva Made in Italy Tra Vulnerabilità ed Eccellenza. Bachelor's Thesis, Università Ca'Foscari, Venezia, Italy, 2016.
8. Cappelli, L.; D'ascenzo, F.; Ruggieri, R.; Rossetti, F.; Scalingi, A. The attitude of consumers towards "Made in Italy" products. An empirical analysis among Italian customers. *Manag. Mark. Chall. Knowl. Soc.* **2019**, *14*, 31–47. [CrossRef]
9. Violino, S.; Antonucci, F.; Pallottino, F.; Cecchini, C.; Figorilli, S.; Costa, C. Food traceability: A term map analysis basic review. *Eur. Food Res. Technol.* **2019**, 1–11. [CrossRef]
10. European Commission. Regulation (EC) No 178/2002 of the European parliament and of the council. Laying down the general principles and requirements of food law, establishing the European Food Safety Authority and laying down procedures in matters of food safety. *Off. J. Eur. Union* **2002**, *31*, 1–24.
11. Violino, S.; Pallottino, F.; Sperandio, G.; Figorilli, S.; Ortenzi, L.; Tocci, F.; Vasta, S.; Imperi, G.; Costa, C. A full technological traceability system for extra virgin olive oil. *Foods* **2020**, *9*, 624. [CrossRef]
12. Antonucci, F.; Figorilli, S.; Costa, C.; Pallottino, F.; Raso, L.; Menesatti, P. A review on blockchain applications in the agri-food sector. *J. Sci. Food Agric.* **2019**, *99*, 6129–6138. [CrossRef] [PubMed]

13. Espiñeira, M.; Santaclara, F.J. What is food traceability. In *Advances in Food Traceability Techniques and Technologies*; Woodhead Publishing: Cambridge, UK, 2016; pp. 3–8.

14. Perri, E.; Benincasa, C. Olive oil traceability. In *Olive Germplasm: The Olive Cultivation, Table and Olive Oil Industry in Italy*; Muzzalupo, I., Ed.; IntechOpen: London, UK, 2012; pp. 265–286.

15. Ou, G.; Hu, R.; Zhang, L.; Li, P.; Luo, X.; Zhang, Z. Advanced detection methods for traceability of origin and authenticity of olive oils. *Anal. Methods* **2015**, *7*, 5731–5739. [CrossRef]

16. Aparicio, R.; García-González, D.L. Olive oil characterization and traceability. In *Handbook of Olive Oil*; Springer: Boston, MA, USA, 2013; pp. 431–478.

17. Benincasa, C.; Lewis, J.; Perri, E.; Sindona, G.; Tagarelli, A. Determination of trace element in Italian virgin olive oils and their characterization according to geographical origin by statistical analysis. *Anal. Chim. Acta* **2007**, *585*, 366–370. [CrossRef] [PubMed]

18. Portarena, S.; Farinelli, D.; Lauteri, M.; Famiani, F.; Esti, M.; Brugnoli, E. Stable isotope and fatty acid compositions of monovarietal olive oils: Implications of ripening stage and climate effects as determinants in traceability studies. *Food Control* **2015**, *57*, 129–135. [CrossRef]

19. Benincasa, C.; Romano, E.; Pellegrino, M.; Perri, E. Characterization of phenolic profiles of Italian single cultivar olive leaves (*Olea europaea* L.) by mass spectrometry. *Mass Spectrom. Purif. Tech.* **2018**, *4*, 2. [CrossRef]

20. Mohamed, M.B.; Guasmi, F.; Alia, S.B.; Radhouani, F.; Faghim, J.; Triki, T.; Grati, K.N.; Baffi, C.; Lucini, L.; Benincasa, C. The LC-MS/MS characterization of phenolic compounds in leaves allows classifying olive cultivars grown in South Tunisia. *Biochem. Syst. Ecol.* **2018**, *78*, 84–90. [CrossRef]

21. Rotondo, A.; Mannina, L.; Salvo, A. Multiple Assignment Recovered Analysis (MARA) NMR for a direct food labeling: The case study of olive oils. *Food Anal. Methods* **2019**, *12*, 1238–1245. [CrossRef]

22. Hirri, A.; Gammouh, M.; Gorfti, A.; Kzaiber, F.; Bassbasi, M.; Souhassou, S.; Balouki, A.; Oussama, A. The use of Fourier transform mid infrared (FT–MIR) spectroscopy for detection and estimation of extra virgin olive oil adulteration with old olive oil. *Sky J. Food Sci.* **2015**, *4*, 60–66.

23. Merás, I.D.; Manzano, J.D.; Rodríguez, D.A.; de la Peña, A.M. Detection and quantification of extra virgin olive oil adulteration by means of autofluorescence excitation-emission profiles combined with multi-way classification. *Talanta* **2018**, *178*, 751–762.

24. Kumar, S.; Kahlon, T.; Chaudhary, S. A rapid screening for adulterants in olive oil using DNA barcodes. *Food Chem.* **2011**, *127*, 1335–1341. [CrossRef]

25. Tengstrand, E.; Rosén, J.; Hellenäs, K.E.; Aberg, K.M. A concept study on non-targeted screening for chemical contaminants in food using liquid chromatography-mass spectrometry in combination with a metabolomics approach. *Anal. Bioanal. Chem.* **2013**, *405*, 1237–1243. [CrossRef]

26. Knolhoff, A.M.; Croley, T.R. Non-targeted screening approaches for contaminants and adulterants in food using liquid chromatography hyphenated to high resolution mass spectrometry. *J. Chromatogr. A* **2016**, *1428*, 86–96. [CrossRef] [PubMed]

27. Vlek, C.; Steg, L. Human behavior and environmental sustainability: Problems, driving forces, and research topics. *J. Soc. Issues* **2007**, *63*, 1–19. [CrossRef]

28. Mossoba, M.M.; Azizian, H.; Fardin-Kia, A.R.; Karunathilaka, S.R.; Kramer, J.K.G. First application of newly developed FT-NIR spectroscopic methodology to predict authenticity of extra virgin olive oil retail products in the USA. *Lipids* **2017**, *52*, 443–455. [CrossRef] [PubMed]

29. Jiménez-Carvelo, A.M.; Osorio, M.T.; Koidis, A.; González-Casado, A.; Cuadros-Rodríguez, L. Chemometric classification and quantification of olive oil in blends with any edible vegetable oils using FTIR-ATR and Raman spectroscopy. *LWT Food Sci. Technol.* **2017**, *86*, 174–184. [CrossRef]

30. Nenadis, N.; Tsimidou, M. Perspective of vibrational spectroscopy analytical methods in on-field/official control of olives and virgin olive oil. *Eur. J. Lipid Sci. Technol.* **2017**, *119*. [CrossRef]

31. Karunathilaka, S.R.; Kia, A.R.F.; Srigley, C.; Chung, J.K.; Mossoba, M.M. Nontargeted, rapid screening of extra virgin olive oil products for authenticity using near-infrared spectroscopy in combination with conformity index and multivariate statistical analyses. *J. Food Sci.* **2016**, *81*, 2390–2397. [CrossRef] [PubMed]

32. Cayuela, J.A.; García, J.F. Sorting olive oil based on alpha-tocopherol and total tocopherol content using Near-Infra-Red Spectroscopy (NIRS) analysis. *J. Food Eng.* **2017**, *202*, 79–88. [CrossRef]

33. Costa, A.F.; Coelho, M.J.; Gambarra, F.F.; Bezerra, S.R.; Harrop, R.K.; Ugulino, M.C. NIR spectrometric determination of quality parameters in vegetable oils using PLS and variable selection. *Food Res. Int.* **2008**, *41*, 341–348.

34. Eskin, M.G.; Torabfam, M.; Psillakis, E.; Cincinelli, A.; Kurt, H.; Yüce, M. Real-time water quality monitoring of an artificial lake using a portable, affordable, simple, arduino-based open source sensor. *Environ. Eng.* **2019**, *6*, 7–14. [CrossRef]

35. Beghi, R.; Buratti, S.; Giovenzana, V.; Benedetti, S.; Guidetti, R. Electronic nose and visible-near infrared spectroscopy in fruit and vegetable monitoring. *Rev. Anal. Chem.* **2017**, *36*, 36. [CrossRef]

36. Galtier, O.; Dupuy, N.; Le Dreau, Y.; Ollivier, D.; Pinatec, C.; Kister, J.; Artaud, J. Geographic origins and compositions of virgin olive oils determined by chemometric analysis of NIR spectra. *Anal. Chem. Acta* **2006**, *595*, 136–144. [CrossRef]

37. Azizian, H.; Mossoba, M.M.; Fardin-Kia, A.R.; Delmonte, P.; Karunathilaka, S.R.; Kramer, J.K.G. Novel, rapid identification, and quantification of adulterants in extra virgin olive oil using near-infrared spectroscopy and chemometrics. *Lipids* **2015**, *50*, 705–718. [CrossRef]

38. Langsrud, O. 50–50 Multivariate analysis of variance for collinear responses. *J. R. Stat. Soc. Ser. D* **2002**, *51*, 305–317. [CrossRef]

39. Infantino, A.; Zaccardelli, M.; Costa, C.; Ozkilinc, H.; Habibi, A.; Peever, T. A new disease of grasspea (*Lathyrus sativus*) caused by *Ascochyta lentis* var. *lathyri*. *Int. J. Plant Pathol.* **2016**, 541–548.

40. MacKay, D.J.C. Bayesian interpolation. *Neural Comput.* **1992**, *4*, 415–447. [CrossRef]

41. Foresee, F.D.; Martin, T.H. Gauss-Newton approximation to Bayesian learning. In Proceedings of the International Joint Conference on Neural Networks IEEE, Houston, TX, USA, 12 June 1997; Volume 3, pp. 1930–1935.

42. Kennard, R.W.; Stone, L.A. Computer aided design of experiments. *Technometrics* **1969**, *11*, 137–148. [CrossRef]

43. Banko, M.; Brill, E. Scaling to very large corpora for natural language disambiguation. In Proceedings of the 39th Annual Meeting on Association for Computational Linguistics, Toulouse, France, 6–11 July 2001; pp. 26–33.

44. Bucci, R.; Magrí, A.D.; Magrí, A.L.; Marini, D.; Marini, F. Chemical authentication of extra virgin olive oil varieties by supervised chemometric procedures. *J. Agric. Food Chem.* **2002**, *50*, 413–418. [CrossRef]

45. Portarena, S.; Baldacchini, C.; Brugnoli, E. Geographical discrimination of extra-virgin olive oils from the Italian coasts by combining stable isotope data and carotenoid content within a multivariate analysis. *Food Chem.* **2017**, *215*, 1–6. [CrossRef]

46. Giuliani, A.; Cerretani, L.; Cichelli, A. Chlorophylls in olive and in olive oil: Chemistry and occurrences, critical. *Food Sci. Nutr.* **2011**, *51*, 678–690. [CrossRef]

47. El Sohaimy, A.A.S.; El-Sheikh, H.M.; Refaay, M.T.; Zaytoun, A.M. Effect of harvesting in different ripening stages on olive (*Olea europaea*) oil quality. *Am. J. Food Technol.* **2016**, *11*, 1–11. [CrossRef]

48. Aprile, A.; Negro, C.; Sabella, E.; Luvisi, A.; Nicolì, F.; Nutricati, E.; Vergine, M.; Miceli, A.; Blando, F.; De Bellis, L. Antioxidant activity and anthocyanin contents in olives (*cv.* Cellina di Nardò) during ripening and after fermentation. *Antioxidants* **2019**, *8*, 138. [CrossRef]

49. Mínguez-Mosquera, M.I.; Gandul-Rojas, B.; Garrido-Fernández, J.; Gallardo-Guerrero, L. Pigments present in virgin olive oil. *J. Am. Oil Chem. Soc.* **1990**, *67*, 192–196. [CrossRef]

50. Criado, M.N.; Motilva, M.J.; Goni, M.; Romero, M.P. Comparative study of the effect of the maturation process of olive fruit on the chlorophyll and carotenoid fractions of drupes and virgin olive oils of Arbequina variety in Spain. *Food Chem.* **2007**, *100*, 748–755. [CrossRef]

51. Ramírez-Anaya, J.; Samaniego-Sánchez, C.; Castañeda-Saucedo, M.C.; Villalón-Mir, M.; López-García de la Serrana, H. Phenols and the antioxidant capacity of Mediterranean vegetables prepared with extra virgin olive oil using different domestic cooking techniques. *Food Chem.* **2015**, *188*, 430–438. [CrossRef]

52. Roca, M.; Minguez-Mosquera, M.I. Change in the natural ratio between chlorophylls and carotenoids in olive fruit during processing for virgin olive oil. *J. Am. Oil Chem. Soc.* **2001**, *78*, 133–138. [CrossRef]

53. Schwartz, S.J.; Woo, S.L.; Von Elbe, J.H. High-performance liquid chromatography of chlorophylls and their derivatives in fresh and processed spinach. *J. Agric. Food Chem.* **1981**, *29*, 533–535. [CrossRef]

54. Aitzetmüller, K. Chlorophyll-abbauprodukte in pflanzlichen ölen. *Eur. J. Lipid Sci. Technol.* **1989**, *91*, 99–105. (In German) [CrossRef]

55. Gertz, C.; Fiebig, H.J. Pyropheophytin a—Determination of thermal degradation products of chlorophyll a in virgin olive oil. *Eur. J. Lipid Sci. Technol.* **2006**, *108*, 1062–1065. [CrossRef]

56. Jaswir, I.; Noviendri, D.; Hasrini, R.F.; Octavianti, F. Carotenoids: Sources, medicinal properties and their application in food and nutraceutical industry. *J. Med. Plant Res.* **2011**, *5*, 7119–7131.

57. Harwood, J.; Aparicio, R. (Eds.) *Handbook of Olive Oil: Analysis and Properties*; Springer: New York, NY, USA, 2000.

58. Psomiadou, E.; Tsimidou, M. Pigments in Greek virgin olive oils: Occurrence and levels. *J. Sci. Food Agric.* **2001**, *81*, 640–647. [CrossRef]

59. Moyano, M.J.; Melendez, A.J.; Alba, J.; Heredia, F.J. A comprehensive study on the colour of virgin olive oils and its relationship with their chlorophylls and carotenoids indexes (I): CIEXYZ non-uniform colour space. *Food Res. Int.* **2008**, *41*, 505–512. [CrossRef]

60. Giuffrida, D.; Salvo, F.; Salvo, A.; Cossignani, L.; Dugo, G. Pigments profile in monovarietal virgin olive oils from various Italian olive varieties. *Food Chem.* **2011**, *124*, 1119–1123. [CrossRef]

61. Domenici, V.; Ancora, D.; Cifelli, M.; Serani, A.; Veracini, C.A.; Zandomeneghi, M. Extraction of pigment information from near-UV vis absorption spectra of extra virgin olive oils. *J. Agric. Food Chem.* **2014**, *62*, 9317–9325. [CrossRef]

62. Fuentes, E.; Baez, M.E.; Bravo, M.; Cid, C.; Labra, F. Determination of total phenolic content in olive oil samples by UV-visible spectrometry and multivariate calibration. *Food Anal. Methods* **2012**, *5*, 1311–1319. [CrossRef]

63. Murphy, K.J.; Walker, K.M.; Dyer, K.A.; Bryan, J. Estimation of daily intake of flavonoids and major food sources in middle-aged Australian men and women. *Nutr. Res.* **2019**, *61*, 64–81. [CrossRef]

64. Torres, A.; Espínola, F.; Moya, M.; Alcalá, S.; Vidal, A.M.; Castro, E. Assessment of phenolic compounds in virgin olive oil by response surface methodology with particular focus on flavonoids and lignans. *LWT* **2018**, *90*, 22–30. [CrossRef]

65. Qiao, Y.Q.; Jiang, P.F.; Gao, Y.Z. Lutein prevents osteoarthritis through Nrf2 activation and down regulation of inflammation. *Arch. Med. Sci. AMS* **2018**, *14*, 617–624. [CrossRef]

Article

Alignment and Proficiency of Virgin Olive Oil Sensory Panels: The OLEUM Approach

Sara Barbieri [1], Karolina Brkić Bubola [2], Alessandra Bendini [1,*], Milena Bučar-Miklavčič [3], Florence Lacoste [4], Ummuhan Tibet [5], Ole Winkelmann [6], Diego Luis García-González [7] and Tullia Gallina Toschi [1]

1   Alma Mater Studiorum—Università di Bologna, 40127 Bologna, Italy; sara.barbieri@unibo.it (S.B.); tullia.gallinatoschi@unibo.it (T.G.T.)
2   Institute of Agriculture and Tourism, HR-52440 Poreč, Croatia; karolina@iptpo.hr
3   Science and Research Centre Koper, 6000 Koper, Slovenia; milena.miklavcic@guest.arnes.si
4   Institut des Corps Gras, 33600 Pessac, France; F.LACOSTE@iterg.com
5   Ulusal Zeytin ve Zeytinyağı Konseyi, 35100 Izmir, Turkey; ummuhan.tibet@uzzk.org
6   Eurofins Analytik GmbH, 21029 Hamburg, Germany; OleWinkelmann@eurofins.de
7   Instituto de la Grasa (CSIC), 41013 Sevilla, Spain; dlgarcia@ig.csic.es
*   Correspondence: alessandra.bendini@unibo.it; Tel.: +39-0547-338121

Received: 16 January 2020; Accepted: 15 March 2020; Published: 19 March 2020

Abstract: A set of 334 commercial virgin olive oil (VOO) samples were evaluated by six sensory panels during the H2020 OLEUM project. Sensory data were elaborated with two main objectives: (i) to classify and characterize samples in order to use them for possible correlations with physical–chemical data and (ii) to monitor and improve the performance of panels. After revision of the IOC guidelines in 2018, this work represents the first published attempt to verify some of the recommended quality control tools to increase harmonization among panels. Specifically, a new "decision tree" scheme was developed, and some IOC quality control procedures were applied. The adoption of these tools allowed for reliable classification of 289 of 334 VOOs; for the remaining 45, misalignments between panels of first (on the category, 21 cases) or second type (on the main perceived defect, 24 cases) occurred. In these cases, a "formative reassessment" was necessary. At the end, 329 of 334 VOOs (98.5%) were classified, thus confirming the effectiveness of this approach to achieve a better proficiency. The panels showed good performance, but the need to adopt new reference materials that are stable and reproducible to improve the panel's skills and agreement also emerged.

Keywords: virgin olive oil; quality; sensory analysis; panel test

---

1. Introduction

The sensory methodology for virgin olive oils (VOOs) known as the "panel test" was proposed in 1987 [1] and, to date, represents the most valuable approach to assess sensory characteristics and quality for consumer and producer protection [2]. The purpose of the method is to standardize procedures for evaluation of the organoleptic characteristics of VOOs and to establish specific quality grades (extra virgin olive oil—EV, virgin olive oil—V, ordinary virgin olive oil—O, lampante olive oil—L). A group of assessors selected in a controlled manner, suitably trained to identify and measure the intensity of positive and negative sensations, represents the analytic tool of this methodology. A collection of methods and standards has been adopted by the International Olive Council (IOC) for sensory analysis of VOOs. These documents describe the vocabulary that tasters must adopt, the characteristics that the sensory laboratory must possess, the tasting conditions and characteristics of the glass for organoleptic analysis of oils, and the sensory method and rules for the selection, training, and monitoring of skilled virgin olive oil tasters [3–7].

In 1991, the method was included into European regulations and obtained the legal validity for establishing the quality grade of the product that included only three categories of VOOs: EV, V, and L [8]. Since application of the method showed some drawbacks, drawing on its own experience, the IOC made a series of revisions to render the method simpler and more reliable [9].

In 2002, the most important innovation introduced was the application of a statistical index to classify oils according to the median of the main perceived defect (mpd) and the median of the fruity attribute that represents the most important positive descriptor. A limit for the value of the robust variation coefficient, which must be no greater than 20%, was also established. In fact, the use of statistical procedures to analyze sensory data is fundamental, as it provides reliable results that are required for data from other analytical methods [10]. Subsequent amendments and revisions concerned, for example, a list of sensory defects, the specific optional terminology for labeling purposes or tasting conditions, have been adopted up to now [9].

Although it has been responsible for improving the quality of VOOs in the last 28 years, the Panel test is frequently under scrutiny. The problem mainly focuses on the debated classification of borderline oils (EV vs. V, V vs. L), reproducibility of results among different laboratories, the limited number of samples that can be analyzed per day (four samples in each session with a maximum of three sessions per day) and the presence of at least 8–12 trained individuals for each sensory evaluation [6]. Another problem in applying the panel test method depends on the lack of appropriate reference standards for training assessors [11]. In addition, some recent commentaries [12,13] discussed methodological features that should be performed more accurately in order to avoid disagreements between different panels. To overcome this, some strategies were proposed that should be applied during different steps of training of assessors (determination of the group detection threshold, selective trials) or during official tasting sessions (alternative approach to the CVr%) for overcoming these difficulties.

Organoleptic assessment is both a qualitative and quantitative method since its application results in the classification of samples based on the median of the main predominant defect and the presence or absence of the fruity attribute. Consequently, assessors in each panel must be effectively trained for correct classification of samples and for correct recognition of the intensities of perceived attributes.

In this context, the OLEUM project "Advanced Solutions for Assuring Authenticity and Quality of Olive Oil at Global Scale" funded by the European Commission within the Horizon 2020 Programme (2014–2020, grant agreement No. 635690), is engaged in reinforcing the methodology for sensory evaluation through design of a global procedure named the "quantitative panel test". This approach aims to improve the activity of sensory panels, whose work remains central to ensuring the quality of the product by: (i) reducing the number of samples to be assessed by the sensory panel by establishing chemometric models (calibrated on a large dataset of reliable sensory classified VOOs) that are able to predict assignment of samples to a specific quality grade using rapid instrumental screening methods, which could allow pre-classification with a certain level of probability that can allow the panels to focus more on sensory analysis of uncertain samples; (ii) increasing the panel's performance by introducing new artificial reference materials validated by a number of sensory panels (six in the case of the OLEUM project) and formulated ad hoc to resemble specific sensory attributes (e.g., rancid and winey); (iii) relating attributes and defects found in VOO with specific molecules (volatile compounds) in order to have an additional qualitative and quantitative tool (quantitation of specific volatile compounds) to support the panel test in confirmatory analyses or in cases of disagreement between panels.

In this regard, some recent works deal with the monitoring of the presence of molecular markers related to specific sensory defects in VOO headspace [14] together with the setting up of chemometric models based on volatile compounds for the prediction of sensory characteristics [15].

The present paper does not aim to illustrate the entire scheme and all the methods involved in the "quantitative panel test". However, in the framework of the panel test, it highlights possibilities for amelioration and describes the proficiency improvement given by formative training, and the method used to obtain sensory classified samples from analysis of a set of VOOs to be used for calibration of rapid instrumental screening methods. Herein, the results of the sensory evaluation of

334 samples are reported and discussed with the aim to: (i) verify the effectiveness of application of the sensory method to evaluate the quality of the product according to [8] and latter modifications (EV, V, L); (ii) highlight the importance of the strict application of IOC guidelines for quality control methodology (for selecting, training, and monitoring tasters and panels) by the sensory panels in performing organoleptic assessment of VOO; (iii) obtain the most reliable sensory classification of samples in terms of quality grades and intensities of positive (fruity) and negative (main perceived defect/s) attributes; this can be achieved by application of a newly proposed "decision tree" that includes formative reassessments when misalignments on the category or the main perceived defect and/or fruity attribute occur.

Many studies in the literature have discussed the relationship between the official sensory method applied by a trained panel and consumer perception [16–21], but to our knowledge there are few studies comparing the results of different trained panels and none aimed to reinforce the application of the official method and increase harmonization among different panels. Specifically, the key elements of this work are: (i) the very large dataset obtained by collecting 334 oils from two olive harvest seasons, representative of the most common olive cultivars, different geographical origins, different sensory profiles, and, especially, the main sensory defects perceived; (ii) the processing of data provided by several panels to obtain a reliable classification by the application of a new decision tree useful for possible correlations with instrumental data and/or for building discriminating models by different instrumental approaches.

2. Materials and Methods

2.1. Sensory Panels

Six panels from six different countries were involved in the sensory analysis carried out in the OLEUM project: EUROFINS from Germany, coded as EU; IPTPO from Croatia, coded as IP; ITERG from France, coded as IT; UNIBO from Italy, coded as UN; UP/ZRS from Slovenia, coded as UP/ZRS; and UZZK from Turkey, coded as UZ.

Each panel has some sort of public authority recognition (national authorities; International Olive Council, IOC; national accreditation bodies for EU standards) [22] and takes part in national and international interlaboratory proficiency tests (organized by private or public authorities) and/or IOC interlaboratory comparison. Their sensory activities are focused on evaluation of the grade of quality (quality control), PDO/PGI certification, olive oil competition, and sensory analysis of samples involved in scientific research. The number of samples evaluated each year by the six panels varies from 125 to 1800. The UNIBO panel was responsible for coordinating the activities of panels and for elaboration of sensory data.

2.2. VOO Samples

Each sensory panel (EU, IT, IP, UN, UP/ZRS, UZ) was responsible for the sampling (two years: 2016–2017 and 2017–2018 olive harvest seasons) of a possibly balanced number of extra virgin (EV), virgin (V), and lampante (L) samples defined by sensory evaluation, according to EU Regulations [8] and later modifications.

These samples were collected to be representative of the most common olive cultivars, different geographical origins (without restrictions to the national market of each sensory panel), different sensory profiles and, especially, the main sensory defects perceived. Samples were directly requested from olive oil companies under a nondisclosure agreement containing information related to responsibility and confidentiality of data. The selection of the sample set for each year was based on sensory screening: each panel leader, assisted by his/her deputy panel leader, was responsible for applying the official procedure for assessing the organoleptic characteristics of VOOs (according to [8] and later modifications). At the end of the sensory screening, panel leaders sent the results of sample screening to the UNIBO panel leader.

Each olive oil company that agreed to participate in the sampling phase had to provide information related to each commercial sample furnished (date of sampling, geographical origin of olives, olive variety/varieties, PDO, PGI, sanitary state of olives, time and storage conditions of olives before milling, mill location, technology parameters, date of production of the oil, date of start of oil storage, type of storage tank/bottles, oil storage temperature). The need to collect all available information on each sample was the reason why samples were requested from companies, avoiding collection directly on the market. However, the oils collected were representative of possible commercial samples and also so-called borderline samples that can be the object of disagreement between panels in terms of sensory characteristics. The olive oil company indicated a person who was responsible for oil sampling among its employees, who followed and applied guidelines for oil sampling [23,24].

For each sample, a volume of 7 L was requested from the olive oil company and collected inside adequate tins or bottles. In case of a batch produced and packed (oil already bottled), the responsible person selected by the olive oil company had to collect the volume required (7 L) taken with a random selection of bottles. The panel leader assisted by his/her deputy panel leader was responsible for managing the laboratory samples. The olive oil company dispatched the packaged samples to the panel leader, who organized the preparation of laboratory samples (after proper homogenization, using a 0.5 L tin), their label codes, and shipment to all the OLEUM partners involved (carried out in the shortest time possible by tracked shipments). Sample codes summarized the basic information: partner acronym (responsible for the sampling) and number (progressive for two years, related in unique way for each sample).

The samples collected for each year (first year, 180 oils; second year, 154 oils) were divided into four subgroups and their sensory evaluation as well as sensory results were planned over time by the UNIBO panel. All samples, stored in the lab at 10–12 °C, were reconditioned at room temperature for 6–8 h before preparing samples for sensory analysis.

2.3. Sensory Analysis

The panel test method was carried out using six OLEUM panels. Positive and negative descriptors were evaluated according to the official procedure ([8] and later modifications). The intensity of each attribute was graded by assessors using a continuous unstructured line scale of 10 cm. Each 15 mL sample was tasted at 28 ± 2 °C in a tasting booth, regulated in terms of shape and equipment [4]. Each panel leader collected the profile sheets completed by each taster (8–12) from his/her panel, reviewed the intensities assigned to the different attributes, and inserted the sensory data in the IOC Excel program for statistical elaboration based on the calculation of the median. The robust coefficients of variation (CVr%) were calculated and validated ([8] and later modifications).

Moreover, with the aim to monitor and possibly improve the performance of panels, after elaboration of each subgroup of sample data, the UNIBO panel, being responsible for the sensory activities, adopted and applied the quality control procedures to check the validity of the results obtained by OLEUM panels in agreement with IOC guidelines for the quality control of virgin olive oil panels revised in 2018 [25], specifically: (i) z-score estimation was conducted for each sensory panel (IOC z-score and OLEUM z-score) to estimate the panel's trueness; (ii) available IOC standards and other materials characterized (samples from previous IOC proficiency tests) were provided to each panel for training purposes; (iii) replicate analysis of three samples selected between the entire set of samples, to estimate panel precision, was performed.

2.4. Statistical Analysis

For processing sensory data from the assessors of each panel (by each respective panel leader), the IOC excel spreadsheet was applied according to official methodology ([6,8] and later modifications). Sensory data from each panel were processed (by the UNIBO panel leader), and after application of the proposed decision tree, the coefficient of variation (CV), was calculated [26] (dataset). A limit for the CV based on its frequency distribution was also proposed to check the level of variability.

The CV frequency distribution was also expressed as cumulative probability by the *t*-test (Student's test distribution).

For control of the performance of the panel, estimation of both precision and trueness of panels was performed according to IOC guidelines [25]. The estimation of the precision of panels was made during the procedure of replicate analysis by the calculation of both normalized error (En) and repeatability number (rN), whereas control of panel trueness was obtained by z-score estimation.

3. Results and Discussion

3.1. The Decision Tree

For the "quantitative panel test", it was very important to classify samples to reach agreement (among the six panels involved) on sensory characteristics (in terms of intensity of positive/negative attributes), thus providing useful information for instrumental analysis.

For this specific objective, the classification of samples based on the evaluation data provided by the six panels was elaborated by applying a decision tree (Figure 1), a new tool for categorization of VOOs.

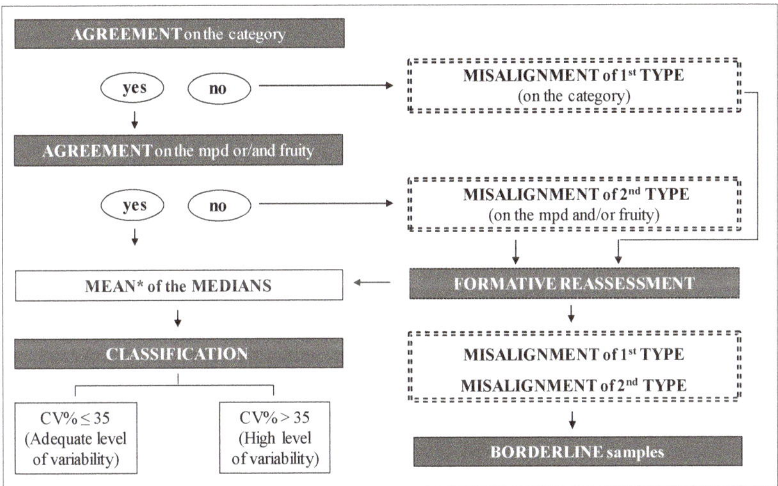

Figure 1. Decision tree adopted for statistical processing of sensory results provided by the six panels. mpd = main perceived defect. * Mean value calculated on the median values obtained by OLEUM panels for mpd and the fruity attribute.

The adopted decision tree is based on the agreement (more than 50% of panels) on the category and on the median of the intensity of the main perceived defect (mpd) and/or of the fruity attribute. If one of these two agreements was not reached, a first or second type of misalignment occurred, and the sample was not classified. Following the flow of the decision tree, the UNIBO panel leader first checked whether the sensory data provided by at least four out of six panels defined the sample as belonging to the same quality grade; if yes, agreement on the mpd was also checked, while in the negative case, formative reassessment was required.

If the desired agreement was met for both criteria, it was possible to proceed with calculation of the mean of the medians (provided by each panel) for classifying the samples. The coefficient of variation (CV%) was applied and considered satisfactory if ≤35% (adequate level of variability). The adoption of 35% as upper limit of CV was selected by observing the frequency distribution of all CV% values registered for the mpd and fruity attribute for the set of samples analyzed. The frequency distribution was also expressed as cumulative probability ($p = 0.74$) applying the *t*-test (Figure 2).

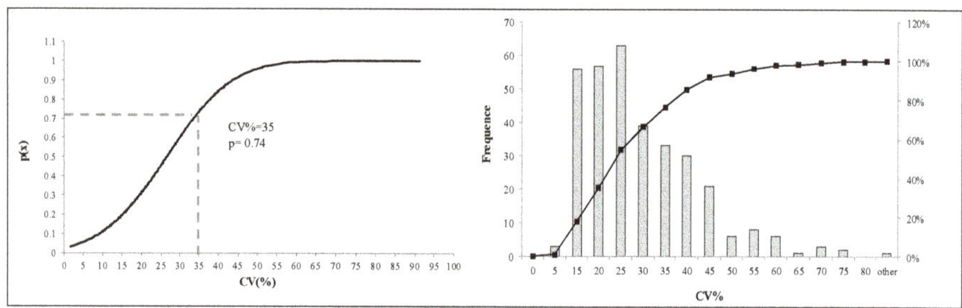

Figure 2. Control of the level of variability of values obtained by application of the decision tree based on the frequency distribution of CV%. CV% = variability of the median values with respect to the mean value. The frequency distribution was also expressed as cumulative probability by *t*-test (Student's test distribution).

Cerretani and co-workers investigated the relationship between sensory and chemical composition of VOOs to assess correlations between sensory attributes and minor components [27]; in this study, sensory attributes were assessed by four panels (two Italian and two Spanish) employing a total of 59 tasters, and the median values for each VOO evaluated by panels were used as the final input for statistical analysis. In our work, the mean of the medians provided by each panel was considered. The median represents the midpoint of an ordered set of odd numbers or the mean of two midpoints of an ordered set of even numbers. It is, therefore, a robust tool since it is not influenced by outliers; considering that it was already applied by each panel individually, the mean of the medians was also considered more appropriate for comparison of results of panels and for monitoring performance.

The decision tree was applied to the entire set of 334 oils and, in case of misalignments, samples were reassessed in a sensory session (formative reassessment) where each panel was provided with the available IOC reference materials and certified oils evaluated by at least three accredited panels (sent by the UNIBO panel) to improve the identification of any defects and assessment of their intensity. The reassessments were done in a blind way (no information related to the type of misalignments were provided to panel leaders), again applying the organoleptic assessment method, but without open discussion of the attributes between assessors.

During the first year of the project, 176 of 180 oils were classified, and only four misalignments occurred (Table S1a–d); in summary, 152 of 180 samples were immediately classified, and 28 samples were reassessed since first- and/or second-type misalignments occurred (14 samples for each type of misalignment). At the end of formative reassessment, 176 samples were classified (54 EV, 76 V, and 48 L), but classification was not possible for four samples (UN_10, UP_14, EU_29, and UN_32) since agreement among four of six panels was not reached. Specifically, disagreement on the category (V/L) was obtained for UN_10 and UP_14, but for both, fusty-muddy sediment and rancid were perceived by at least four of six panels, indicating these samples as representative of borderline samples; on the other hand, for samples EU_29 and UN_32, an agreement on the category (V) was reached, but not on the identity of mpd due to the presence of more than one defect (fusty-muddy sediment, musty, winey, frostbitten olives, rancid were indicated for EU_29; fusty-muddy sediment, frostbitten olives, rancid were indicated for UN_32), but none were perceived by at least 50% of the panels.

The sensory evaluation of oils from the second sampling (2017/2018 oil campaign), as well as the application of the decision tree, allowed the classification of this set (154 oils) as follows: 69 classified as EV, 51 classified as V, 33 classified as L; one sample was not classified due to an anomalous lemon smell (ZRS_1) and was therefore excluded from the set (Table S2a–d). For 17/154 oils, misalignments of first or second type were achieved (15 and 2, respectively) but, after formative reassessment, all samples were classified by OLEUM panels.

A recent comparative study [28] on a panel test made by nine IOC recognized panels (five from Italy, two from Spain, one from Greece, and one from Slovenia) and chemical analysis of commercial olive oils (16 samples) reported that the sensory methodology works well in case of extremely good olive oils, but not for common commercial ones, and therefore it should be applied only for Protected Designation of Origin (PDO) and other peculiar EVs. Results from the present work, carried out on a large set of commercial VOOs, are in disagreement with those of Circi et al. [28]. The panel test is an official method that has been used to assess improvement in the quality of VOOs since 1991 up to now and provides information on sensory characteristics (intensities of fruity, bitter, and pungent; presence of more than one defect) that are difficult to obtain using a single instrumental approach. The strict application of IOC guidelines for training and quality control of panels and some improvements in the training of a sensory panel, such as the availability of new reference materials that are stable and reproducible, is crucial to increase the reliability of a method to apply a group of assessors as an analytic tool.

3.2. The Panel's Performance

The UNIBO panel, responsible for statistical elaboration of the sensory results, in agreement with the guidelines of IOC document T.28 revised in 2018 [25], summarized the z-score (satisfactory, questionable or unsatisfactory results) for each subgroup of samples from each year and sent it to panel leaders to help them in monitoring the performance of their own panel and to adopt any corrective actions.

The same method adopted by the IOC during its proficiency test (IOC z-score) was applied; it was calculated using: (i) the median (Me) of the predominant defect (the intensity of predominant defect was considered regardless the type of defect that could be different between the six panels) and/or the fruity attribute detected by each panel; (ii) the great median (assigned value, GM) calculated as median of the medians for the predominant defect or for the fruity attribute (detected by all panels as consensus value); (iii) the standard deviation (σ obj) of the scores calculated from IOC historical data (±0.7). A slightly modified version of this method (OLEUM z-score) was also adopted; the only difference from the previous one was, in case of V and L categories, the use of the median (Me) of the defect identified as predominant by consensus of the panels (even if it was not the predominant defect for each panel).

Therefore, the intensity, and also the type of the mpd, was considered in the OLEUM version of the z-score to obtain a reliable dataset for comparison with instrumental data (e.g., in OLEUM for developing screening methods based on the analysis of volatile compounds). The detailed formulas of both the methods used to calculate the z-score are shown in Figure 3.

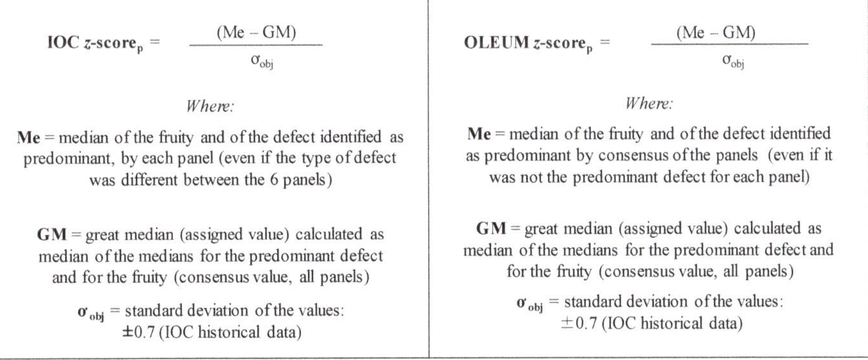

Figure 3. Formulas of the two methods used to calculate the z-score (IOC and OLEUM).

Results of the z-score estimation were illustrated by quality control charts, as part of internal quality control. Some examples of panel performance evaluation are reported in Figures 4 and 5; the vertical axis represents the z-score and the horizontal one identifies the sample codes.

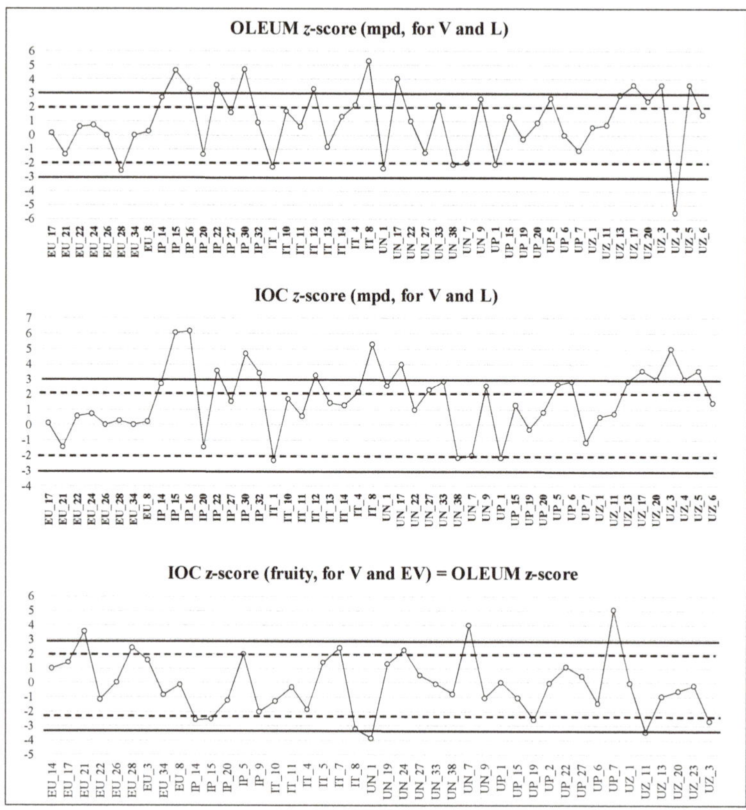

Figure 4. Example of z-score graph for estimation of panel performance, calculated on 60 samples from the subgroup of the first sampling year (180 samples). Criteria of acceptance: |z| ≤ 2, performance was satisfactory; 2 < |z| ≤ 3, performance was questionable; |z| > 3, performance was considered unsatisfactory. The z-scores were calculated for median of the main perceived defect (for V and L category) and for the median of fruity attribute (for V and EV category).

The z-score has positive or negative values and was calculated for both fruity (for EV and V category) and negative sensory attributes (for V and L category); the central value is zero, the warning limits for the index are ±2, and the action limits are ±3. The interpretation is the same for both the methods applied (IOC and OLEUM): if |z| ≤ 2, performance was satisfactory; if 2 < |z| ≤ 3, performance was questionable; finally, if |z| > 3, performance was considered unsatisfactory. Each panel leader, observing this chart, had to define any corrective or/and preventive actions taken if a result is outside of the limits or if several consecutive results are obtained at the same side (positive or negative) of the central value (bias) [25]. The results obtained verified that the approach using the z-score represents a very useful tool to evaluate the trueness of the panel over time.

An example of panel performance reported in Figure 4 showed that, in the case of OLEUM z-score for the mpd (V and L), the panel obtained 25 of 48 satisfactory results, 12 questionable, and 11 unsatisfactory, whereas in the case of IOC z-score, 23 of 48 satisfactory, 14 questionable and 11 unsatisfactory results were obtained. In the case of IOC z-score for fruity attribute (V and EV), the panel obtained 29 of 42 satisfactory results, 7 questionable, and 6 unsatisfactory. These results highlight a trend of the panel to more frequently use higher values of the scale for the intensity of mpd or fruity attribute than the GM value (median of the medians of six panels); moreover, in some cases, the presence of a z-score lower than -2 indicated the lack of intensity recognition of the mpd or of the fruity attribute. The second example (Figure 5) showed that for the mpd (V and L), the panel obtained 18 of 19 satisfactory results and 1 questionable result for OLEUM z-score, while obtaining 17 of 19 satisfactory results and 2 questionable results for IOC z-score.

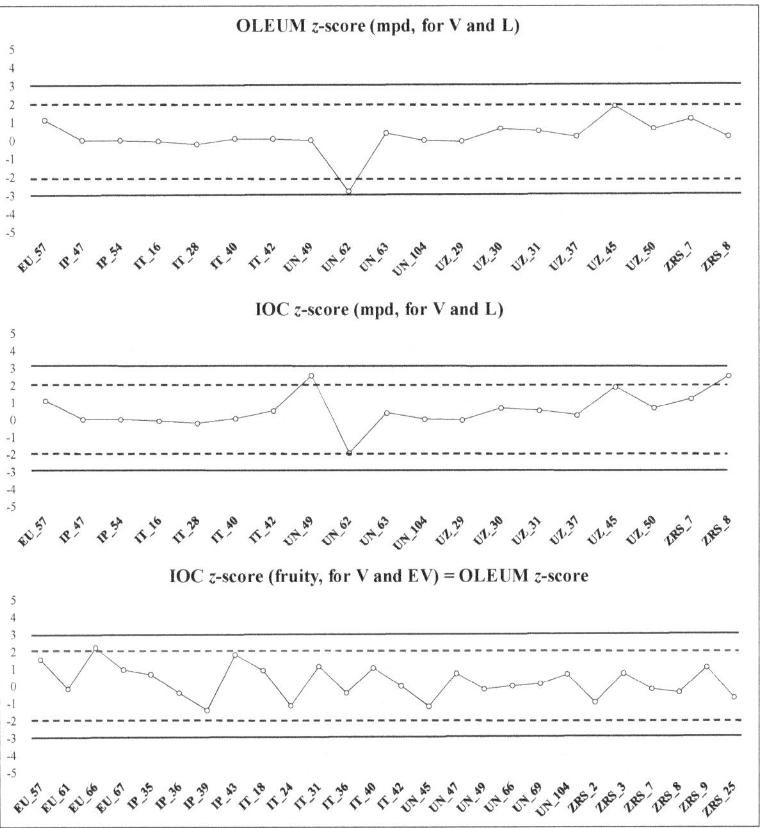

Figure 5. Example of z-score graph for estimation of panel performance, calculated on 38 samples from the third subgroup of the second sampling year (154 samples). Criteria of acceptance: |z| ≤ 2, performance was satisfactory; 2 < |z| ≤ 3, performance was questionable; |z| > 3, performance was considered unsatisfactory. The z-scores were calculated for median of the main perceived defect (for V and L category) and median of fruity attribute (for V and EV category).

In the case of IOC z-score for the fruity attribute (V and EV), the panel obtained 25 of 26 satisfactory results and 1 questionable result. Overall, the panel showed good performance, although the verification of samples in which the z-score is questionable, using both the panel results and those provided by all panels (by the application of the decisional tree), was suggested in the feedback sent to the panel leader. The estimation of z-score was consistent in evaluating the performance of sensory laboratories over time. Its application in this study showed a progressive, greater convergence of results passing from the first to the second sampling and allowed identification of the critical aspects of the performance of each panel and definition of suitable actions for improvement.

In addition to the z-score estimation, during the second year of sampling, the control of the panel's precision was also performed by using replicate analysis. The repeatability of panels was controlled by comparing the medians obtained on three samples in duplicate and determining whether the results are homogenous and, therefore, statistically acceptable.

Specifically, three pairs of identical samples were sent to the panels with different codes (blind conditions) (UN_44 = UN_55, UN_59 = UN_60, and UN_66 = UN_69) and the level of agreement between intensity values expressed for the same sample during independent evaluations was estimated by calculating the repeatability number (rN) and normalized error (En), whose acceptability limits are ≤2 and ≤1, respectively [25] (Table 1).

Table 1. Values of repeatability number (rN), normalized error (En) of each panel for the predominant defect (d) or fruity attribute (f) and suggested limits for these parameters, calculated on the three pairs of samples (UN_44/UN_55, UN_59/UN_60, UN_66/UN_69) evaluated in the replicate analysis (blind conditions).

Panels	UN_44 = UN_55		UN_59 = UN_60		UN_66 = UN_69	
	En_d	rN_d	En_d	rN_d	En_f	rN_f
1	0.3	0.4	0.3	1.3	2.0	14.4
2	0.3	0.4	1.2	5.3	0.6	1.4
3	0.2	0.1	1.2	5.1	0.7	2.0
4	0.1	0.1	0.5	1.1	0.6	1.2
5	0.3	0.3	0.4	0.6	0.7	2.0
6	0.1	0.1	1.2	5.8	0.1	0.0
Limits	≤1	≤2	≤1	≤2	≤1	≤2

In general, the panels showed good repeatability. In the case of the first pair of samples (UN_44 = UN_55, category V), in fact, the values of both parameters (En and rN) were below the suggested limit for good performance; for the second pair of samples (UN_59 = UN_60, category V), the least satisfactory performances were achieved: three panels showed values above these limits (2, 3 and 6), highlighting the need for additional training to improve performance. Finally, for the third replicated sample (UN_66 = UN_69, category EV), only one panel registered values above the limits due to different intensity of the fruity attribute in the two sessions and therefore was not considered repeatable. These indices are based on the evaluation of the correct intensity of the mpd or fruity attribute (and therefore the product quality grade) by each panel and do not take into account the type of defect; results from the application of the decisional tree were consistent for the correct classification of samples, but not for the mpd (UN_44 fusty-muddy sediment, UN_55 rancid, UN_59 brine, UN_60 winey). The inconsistency in the nature of mpd was probably due to more than one defect present in the sample and with similar intensities; in addition, brine and winey usually go together.

4. Conclusions

This work aimed to reinforce the methodology for sensory analysis of VOOs through adoption of supporting tools for training and monitoring of sensory panels. The results obtained from the sensory evaluation carried out by the six panels involved in the OLEUM project on a set of 334 samples confirmed the effectiveness of the application of the panel test. However, at the same time, it also confirmed that there are some critical issues related to questionable results in the case of: (i) borderline oils (between two product categories); or (ii) misalignments on the main perceived defect by panels when more than one negative attribute was present in the oil. The adoption of a decision tree based on the agreement of a category, main perceived defect, and application of formative reassessment in case of misalignments using the same reference materials (samples already classified by the six panels with an high agreement) allowed for reliable classification of oils that, at first evaluation, were borderline. Only 45 of 334 oils were reassessed (formative reassessment) and 41 of 45 samples were definitively classified, confirming the importance of alignment between panels, which can be achieved by sharing the same sensory reference materials. In fact, sensory information on both quality grades of samples and main perceived defect/s is fundamental for testing possible correlations with physical–chemical data and/or for building classification models; in this way, instrumental screening approaches can allow for a reduction in the number of samples that have to be assessed by panels, excluding, for example, oils defintely classified by chemometric models as extra virgin or lampante, focusing the sensory analysis on samples that are not classified or classified with a low probability. This thus reduces the number of samples to be assessed by the sensory panel. The data provided by the panels were also used to verify performance in terms of discriminating capacity, agreement between panels, and accuracy of results by applying some of the procedures reported in the IOC guide for internal quality control of sensory laboratories. In general, the panels showed very good (sometimes excellent) performance even if, in some cases, problems were noted that were related to the use of the scales, lack of recognition of some sensory defects, or intensity values that were too distant between panels for the same sample, especially in the case of oils in which more than one defect was perceived. The large set of samples evaluated over 2 years allowed estimation of the performance of the panel test: the utility and peculiarity of this official method is undisputed, also considering that it has definitively improved the quality of VOOs over the last 28 years, opening the possibility to have a wide range of excellent oils with a deserved added value on the market. On the other hand, to improve its effectiveness, it is necessary that the sensory panels perform organoleptic evaluation by applying specific guidelines [6] and quality control of panel performance [25] in a rigorous manner. To enhance panel skills in recognizing, identifying, and quantifying sensory attributes, the use of new reliable reference materials is of absolute necessity. They could be both "synthetic", resembling a single negative attribute (e.g., rancid or viney-winegary) or biotechnologically formulated, in the latter case being closer to actual virgin olive oils. The first type could be used to overcome some of the limitations of the natural matrix and offer advantages such as feasible preparation in each laboratory (open access composition), reproducibility over time, possibility of purchase, and therefore diffusion and availability for the global market. Even the cultural aspects related with knowledge of the sensory aspects of VOOs, i.e., the global recognition of its positive/negative attributes, could also be facilitated by the availability of these "simplified" materials; the formulation and validation of two of these "synthetic" sensory reference materials (rancid and winey-vinegary) are still in progress within the framework of the OLEUM project. On the other hand, the use of the OLEUM decision tree could be an adequate instrument to classify natural sensory reference materials, for example, those obtained by biotechnological processes (programed fermentations for fermentative defects) or oxidation (for the nonfermentative rancid defect), the availability of which is also fundamental to achieve alignment between panels, thus reducing cases of discordant classifications, which is of vital importance for global trade and product reputation.

Supplementary Materials: The following are available online at http://www.mdpi.com/2304-8158/9/3/355/s1, Table S1 (a–d): Sensory results of samples from the first year, Table S2 (a–d): Sensory results of samples from the second year.

Author Contributions: Conceptualization, S.B., A.B. and T.G.T.; formal analysis, K.B.B., M.B.-M., F.L., U.T. and O.W.; data curation, S.B. and A.B.; writing—original draft preparation, S.B.; writing—review and editing, S.B., A.B., K.B.B., M.B.-M. and T.G.T.; supervision, T.G.T. and D.L.G.-G.; project administration, T.G.T.; funding acquisition, T.G.T. All authors have read and agree to the published version of the manuscript.

Funding: This work is supported by the Horizon 2020 European Research project OLEUM "Advanced solutions for assuring the authenticity and quality of olive oil at a global scale", which received funding from the European Commission within the Horizon 2020 Programme (2014–2020), grant agreement No. 635690.

Acknowledgments: The information expressed in this article reflects the authors' views; the European Commission is not liable for the information contained herein. We are grateful to all producers who provide us with virgin olive oils for this study as well as all panel members who performed sensory analysis of virgin olive oils from each institution involved: Eurofins Analytik GmbH, Hamburg, Germany; Institute of Agriculture and Tourism, Poreč, Croatia; Institut des Corps Gras, Pessac, France; Alma Mater Studiorum—Università di Bologna; Science and Research Centre Koper, Slovenia and Ulusal Zeytin ve Zeytinyağı Konseyi, Izmir, Turkey.

Conflicts of Interest: The authors declare no conflict of interest. The funders had no role in the design of the study; in the collection, analyses, or interpretation of data; in the writing of the manuscript, or in the decision to publish the results.

References

1. International Olive Oil Council. *Sensory Analysis of Olive Oil Method for the Organoleptic Assessment of Virgin Olive Oil*; IOOC/T.20/Doc.no.3; IOOC: Madrid, Spain, 1987.
2. Fernandes, G.D.; Ellis, A.C.; Gámbaro, A.; Barrera-Arellano, D. Sensory evaluation of high-quality virgin olive oil: Panel analysis versus consumer perception. *Curr. Opin. Food Sci.* **2018**, *21*, 66–71. [CrossRef]
3. International Olive Oil Council. *Sensory Analysis. General Basic Vocabulary*; IOC/T.20/Doc.no.4/Rev.1; IOC: Madrid, Spain, 2007.
4. International Olive Council. *Sensory Analysis of Olive Oil Standard, Glass for Oil Tasting*; IOC/T.20/Doc.no.5/Rev.1; IOC: Madrid, Spain, 2007.
5. International Olive Council. *Guide for the Installation of a Test Room*; IOC/T.20/Doc.no. 6/rev.1; IOC: Madrid, Spain, 2007.
6. International Olive Council. *Sensory Analysis of Olive Oil Method for the Organoleptic Assessment of Virgin Olive Oil*; IOC/T.20/Doc. no.15/Rev.10; IOC: Madrid, Spain, 2018.
7. International Olive Council. *Guide for the Selection, Training and Monitoring of Skilled Virgin Olive Oil Tasters*; IOC/T.20/Doc. no.14/Rev.5; IOC: Madrid, Spain, 2018.
8. Official Journal of the European Communities. *European Community, Commission Regulation 2568/91 on the Characteristics of Olive Oil and Olive Residue Oil and on the Relevant Methods of Analysis*; Official Journal of the European Communities: Brussels, Belgium, 1991; Volume L248, pp. 1–83.
9. Conte, L.; Bendini, A.; Valli, E.; Lucci, P.; Moret, S.; Maquet, A.; Lacoste, F.; Brereton, P.; García-González, D.L.; Moreda, W.; et al. Olive oil quality and authenticity: A review of current EU legislation, standards, relevant methods of analyses, their drawbacks and recommendations for the future. *Trends Food Sci. Technol.* **2019**, in press. [CrossRef]
10. Official Journal of the European Communities. *European Community, Commission Regulation 796/2002. Amending Regulation (EEC) No 2568/91No 2568/91/EEC*; Official Journal of the European Communities: Brussels, Belgium, 2002; Volume L128, pp. 8–28.
11. Procida, G.; Giomo, A.; Cichelli, A.; Conte, L.S. Study of volatile compounds of defective virgin olive oils and sensory evaluation: A chemometric approach. *J. Sci. Food Agric.* **2005**, *85*, 2175–2183. [CrossRef]
12. Amelio, M. The official method for olive oil sensory evaluation: An expository revision of certain sections of the method and a viable means for confirming the attribute intensities. *Trends Food Sci. Technol.* **2016**, *47*, 64–68. [CrossRef]
13. Amelio, M. Olive oil sensory evaluation: An alternative to the robust coefficient of variation (CVr%) for measuring panel group performance in official tasting sessions. *Trends Food Sci. Technol.* **2019**, *88*, 567–570. [CrossRef]
14. Romero, I.; García-González, D.L.; Aparicio-Ruiz, R.; Morales, M.T. Validation of SPME–GCMS method for the analysis of virgin olive oil volatiles responsible for sensory defects. *Talanta* **2015**, *134*, 394–401. [CrossRef] [PubMed]
15. Quintanilla-Casas, B.; Bustamante, J.; Guardiola, F.; García-GonzáLez, D.L.; Barbieri, S.; Bendini, A.; Gallina Toschi, T.; Vichi, S.; Tres, A. Virgin olive oil volatile fingerprint and chemometrics: Towards an instrumental screening tool to grade the sensory quality. *LWT Food Sci. Technol.* **2020**, *121*, 1–8. [CrossRef]
16. Delgado, C.; Guinard, J.X. How do consumer hedonic ratings for extra virgin olive oil relate to quality ratings by experts and descriptive analysis ratings? *Food Qual. Prefer.* **2011**, *22*, 213–225. [CrossRef]
17. Recchia, A.; Monteleone, E.; Tuorila, H. Responses to extra virgin olive oils in consumers with varying commitment to oils. *Food Qual. Prefer.* **2012**, *24*, 153–161. [CrossRef]
18. Gámbaro, A.; Ellis, A.; Raggio, L. Virgin olive oil acceptability in emerging olive oil-producing countries. *Food Nutr. Sci.* **2013**, *4*, 1060–1068. [CrossRef]
19. Predieri, S.; Medoro, C.; Magli, M.; Gatti, E.; Rotondi, A. Virgin olive oil sensory properties: Comparing trained panel evaluation and consumer preferences. *Food Res. Int.* **2013**, *54*, 2091–2094. [CrossRef]
20. Barbieri, S.; Bendini, A.; Valli, E.; Gallina Toschi, T. Do consumers recognize the positive sensorial attributes of extra virgin olive oils related with their composition? A case study on conventional and organic products. *J. Food Compos. Anal.* **2015**, *44*, 186–195. [CrossRef]

21. Vazquez-Araujo, L.; Adhikari, K.; Chambers, E.T.; Chambers, D.H.; Carbonell-Barrachina, A.A. Cross-cultural perception of six commercial olive oils: A study with Spanish and US consumers. *Food Sci. Technol. Int.* **2015**, *21*, 454–466. [CrossRef] [PubMed]

22. ISO. *General Requirements for the Competence of Testing and Calibration Laboratories*; EN ISO/IEC 17025:2017; ISO: Geneva, Switzerland, 2017.

23. Official Journal of the European Union. *European Union, Commission Regulation 1348/2013. Amending Regulation (EEC) No 2568/91*; Official Journal of the European Union: Brussels, Belgium, 2013; Volume L338, pp. 31–67.

24. ISO. *Animal and Vegetable Fats and Oils—Sampling*; EN ISO 5555:2001; ISO: Geneva, Switzerland, 2001.

25. International Olive Council. *Guidelines for the Accomplishment of the Requirements of the Norm ISO 17025 by of Sensory Testing Laboratories with Particular Reference to Virgin Olive Oil*; IOC/T.28/Doc.no.1/Rev.4; IOC: Madrid, Spain, 2018.

26. Barbieri, S.; Brkić Bubola, K.; Bendini, A.; Bučar-Miklavčič, M.; Lacoste, F.; Tibet, U.; Winkelmann, O.; García-González, D.L.; Gallina Toschi, T. *OLEUM Project. Sensory Panels Analysis of Commercial Virgin Olive Oil Samples. Results from the First and Second Year of the OLEUM Project*; University of Bologna: Bologna, Italy, 2020. [CrossRef]

27. Cerretani, L.; Desamparados Salvador, M.; Bendini, A.; Fregapane, G. Relationship Between Sensory Evaluation Performed by Italian and Spanish Official Panels and Volatile and Phenolic Profiles of Virgin Olive Oils. *Chemosens* **2008**, *1*, 258–267. [CrossRef]

28. Circi, S.; Capitani, D.; Randazzo, A.; Ingallina, C.; Mannina, L.; Sobolev, A.P. Panel test and chemical analyses of commercial olive oils: A comparative study. *Chem. Biol. Technol. Agric.* **2017**, *4*, 1–10. [CrossRef]

 *foods*

Article

Non-Targeted Authentication Approach for Extra Virgin Olive Oil

Didem Peren Aykas [1,2], Ayse Demet Karaman [3], Burcu Keser [4]  and Luis Rodriguez-Saona [1,*]

[1]   Department of Food Science and Technology, The Ohio State University, 100 Parker Food Science and Technology Building, 2015 Fyffe Road, Columbus, OH 43210, USA; aykas.1@osu.edu

[2]   Department of Food Engineering, Faculty of Engineering, Adnan Menderes University, Aydin 09100, Turkey

[3]   Department of Dairy Technology, Faculty of Agricultural Engineering, Adnan Menderes University, Aydin 09100, Turkey; demet.karaman@adu.edu.tr

[4]   Kocarli Vocational School, Adnan Menderes University, Aydin 09100, Turkey; bkeser@adu.edu.tr

*   Correspondence: rodriguez-saona.1@osu.edu; Tel.: +1-614-292-3339

Received: 28 January 2020; Accepted: 14 February 2020; Published: 20 February 2020

Abstract: The aim of this study is to develop a non-targeted approach for the authentication of extra virgin olive oil (EVOO) using vibrational spectroscopy signatures combined with pattern recognition analysis. Olive oil samples ($n = 151$) were grouped as EVOO, virgin olive oil (VOO)/olive oil (OO), and EVOO adulterated with vegetable oils. Spectral data was collected using a compact benchtop Raman (1064 nm) and a portable ATR-IR (5-reflections) units. Oils were characterized by their fatty acid profile, free fatty acids (FFA), peroxide value (PV), pyropheophytins (PPP), and total polar compounds (TPC) through the official methods. The soft independent model of class analogy analysis using ATR-IR spectra showed excellent sensitivity (100%) and specificity (89%) for detection of EVOO. Both techniques identified EVOO adulteration with vegetable oils, but Raman showed limited resolution detecting VOO/OO tampering. Partial least squares regression models showed excellent correlation (Rval ≥ 0.92) with reference tests and standard errors of prediction that would allow for quality control applications.

Keywords: authenticity; extra virgin olive oil; Raman; FT-IR

---

1. Introduction

Counterfeiters target high-value products, including those with a strong brand name, deceiving consumers by substituting a high-value product with a less expensive or lower quality alternative. Although most food fraud concerns do not result in a public health or food safety crisis, these acts can lead to severe health hazards, as evidenced by oil fraudulently sold as olive oil that caused an outbreak of a condition known as the toxic oil syndrome, affecting 20,000 people, of which more than 300 died in Spain (1981) due to the ingestion of a food-grade rapeseed oil containing aniline derivatives sold for human consumption by street vendors [1]. To prevent olive oil adulteration, global governmental agencies (e.g., European Commission, United States Department of Agriculture, International Olive Council, Codex Alimentarius, German/Australian Standard, North American Olive Oil Association) have developed different standards to regulate olive oil by establishing a set of physical, chemical, and organoleptic characteristics [2]. A 2013 report by the U.S. International Trade Commission (USITC) indicated that current standards for extra virgin olive oil (EVOO) are widely unenforced leading to adulterated and mislabeled products in the market [3]. Common adulterants of EVOO include lower quality olive oils (refined, pomace, or lampante) or seed oils [4].

Numerous analytical techniques have been proposed to detect and control olive oil adulteration, including Ultraviolet-visible (UV–vis) absorption [5,6], front-face total fluorescence spectroscopy [7], vibrational spectroscopy [8–11], mass spectrometry [12–14], nuclear magnetic resonance [15–20], and techniques such as DNA-based methods [21] and electronic noses [22]. Most methods to detect olive oil adulteration have focused on targeted approaches, providing great selectivity and sensitivity for identification and quantification of pre-defined compounds or classes of compounds, but fail to detect emerging risks from unexpected adulterants [23]. On the other hand, non-targeted screening, which is currently at the heart of metabolomics, focuses on the detection of all compounds in a sample without any prior knowledge of chemical entities which can then be compared with the fingerprint profile of pure reference sample [24].

Advancements in semiconductors have allowed miniaturization and cost reduction of spectrometer components, leading to commercially available portable, handheld, compact, and micro-devices in the industry. Key enabling technologies leading to miniaturized structures have been fostered by developments in Micro Electro Mechanical Systems (MEMS), thin-film filters, solid-state lasers, light-emitting devices (LEDs) and alternative light sources, fiber optic assemblies, and high-performance detector arrays [25]. These devices have been at the forefront of cutting-edge technologies and have become progressively smaller and easier to use. Miniaturized devices can be taken to or placed at/in/on-line points of vulnerability along with complex food supply networks and moved from the confines of the relatively stable and controlled laboratory environment into the potentially more challenging and dynamic environs of the food supply chain (point-and-shoot) [26].

Limited information is reported in the literature regarding the detection of olive oil adulteration using non-targeted classification approaches. Mossoba et al. (2017) evaluated FT-NIR in conjunction with a partial least square analysis to predict EVOO authenticity of 93 samples collected from online and local grocery stores [27]. The authors developed an FT-NIR index based on two carbonyl overtone (5280 cm^{-1} and 5180 cm^{-1}) absorptions and generated partial least squares regression (PLSR) models for four specific oils (refined, high oleic, high linoleic, and palm olein) based on the different fatty acid composition of the potential adulterants in EVOO [27]. FT-IR equipped with an attenuated total reflectance (ATR) accessory and combined with supervised pattern recognition techniques (soft independent modeling of class analogy (SIMCA) and partial least squares discriminant analysis (PLS-DA) have detected adulteration of EVOO with vegetable oils at levels above 10% [28,29]. Jimenez-Carvelo and others (2017) evaluated the use of FTIR-ATR and Raman spectroscopy (785 nm excitation laser) with different chemometric classification methods to detect adulteration of olive oil in blends with vegetable oils [30]. They successfully discriminated olive oils from blends containing over 10% vegetable oils by using PLS-DA and support vector machine-classification (SVM-C) for FT-IR and Raman analysis, respectively. Georgouli and others (2017) assessed the capabilities of a compact FTIR-ATR and a bench-top 1064 nm Raman spectrometers on the detection of EVOO adulteration with hazelnut oil (1–90%) mixtures by using a novel continuous locality preserving projections (CLPP) technique accompanied by a k-nearest neighbors (kNN) algorithm, reporting a classification rate ≥69% [31]. Although these studies have shown the capabilities of vibrational spectroscopy to detect EVOO adulteration with vegetable oils, they have not included lower quality olive oil (refined, lampante, or pomace), and most have been developed using a limited number of olive oil samples coming from restricted varietal origins and geographical areas, which limits their use as global methods to detect adulteration of olive oil (independently of the cultivars) with any edible vegetable oil [2,30].

This study aimed to develop an authentication program for EVOO using vibrational spectroscopy signatures combined with pattern recognition analysis for non-targeted screening of commercial EVOO samples and to generate prediction models for monitoring olive oil quality parameters.

2. Materials and Methods

A total of 151 olive oil samples were used in this study. Samples from Turkey (n = 91) were obtained from Aydin Commodity Exchange Laboratories in Aydin, Turkey, which monitors EVOOs for exportation to different countries. In addition, we included EVOO samples that were kindly provided by the California Olive Oil Council ($n = 20$) and samples purchased from grocery stores that included origins from Italy, Spain, Greece, Turkey, Tunisia, Portugal, and Peru ($n = 40$). Oils were placed in amber glass vials and stored at -18 °C until further analysis to minimize oxidation and any compositional changes.

2.1. Reference Methods

The fatty acid profile was determined using a fatty acid methyl ester (FAME) procedure. Fatty acid esterification was achieved by dissolving 100 μL olive oil sample with 10 mL of hexane in a glass tube, after which 100 μL 2N potassium hydroxide in methanol was added and the mixture was vortexed. An aliquot (1.5 mL) was placed into a microcentrifuge tube and rotated at 13.2 rpm for 5 min, and the solution was transferred into a borosilicate glass vial and stored at -18 °C until further Gas Chromatography (GC) analysis. FAMEs were analyzed using an Agilent 6890 series (Santa Clara, CA, USA) GC, equipped with a flame ionization detector (FID) and an HP G1513A autosampler and a tray. Fatty acids' separation was achieved using HP-88 60 m $\times$ 0.25 mm $\times$ 0.2 μm column (Agilent 112-8867), and helium was used as a carrier gas. The injection volume was 1 μL with a split ratio of 20:1. The oven conditions were 110 °C for 1 min, to 220 °C (5 °C/min) hold for 15 min. The injector temperature was 220 °C, and the detector temperature was 250 °C. Fatty acids were identified by comparing each peak's retention times against reference standards (Supelco® 37 Component FAME Mix, Sigma Aldrich, St. Louis, MO, USA). GC analyses were carried out in duplicate.

2.2. Monitoring EVOO Quality Indices

Olive oil samples were analyzed for peroxide value (PV), free fatty acid (FFA) value, pyropheophytins (PPP), and total polar compound (TPC) tests. PV and FFA of the samples were determined using a Metrohm, 916 Ti-Touch (Herisau, Switzerland) automatic titrator. The PV test was performed using a Metrohm Pt Titrode electrode (Herisau, Switzerland), by following the AOCS official method Cd 8-53 [32] and expressed as meqO$_2$/kg of oil. The FFA test was carried out using a Metrohm Solvotrode electrode (Herisau, Switzerland) and following the European Pharmacopoeia 5.0 01/2005:20501 modifications to the AOCS official method Ca 5a-40 [33]. FFA results were expressed in terms of the percentage of oleic acid. Pyropheophytin analysis was carried out by following the ISO 29841:2009/AMD 1:2016 [34] official method and by using a high-performance liquid chromatography (HPLC) (1100 Series, Agilent Technologies, Santa Clara, CA, USA) that was equipped with a G1311A quaternary pump, a G1322A degasser, a G1313 ALS autosampler, and a G1315B DAD detector (Agilent Technologies, Santa Clara, CA, USA). The separated pheophytin components were monitored at 410 nm. The results were expressed as relative proportions (%) of the analytes (pheophytin a and a', and pyropheophytin a). Total polar compound (TPC) content was determined using Testo 270 oil tester (West Chester, PA, USA), according to the manufacturer's operation guide and expressed as a percentage. All the reference tests were carried out in duplicate.

2.3. Vibrational Spectroscopy

Before the data collection, all the olive oil samples were heated to 65 °C in a lab oven (Precision Standard Incubator, PR205125G, Thermo Fisher Scientific, Waltham, MA, USA) to liquefy all the samples to the same level. FT-IR Spectroscopy: Spectra of each oil sample were acquired using a portable 5500a series compact Fourier-Transform IR spectrometer (Agilent Technologies Inc., Santa Clara, CA, USA) equipped with a temperature controlled, 5-reflections ZnSe crystal attenuated total reflectance (ATR) accessory, which was set to 65 °C to prevent fat solidification during the spectral collection. Thermoelectrically-cooled deuterated triglycine sulfate (dTGS) detector was used to measure the amount of light absorbed by the sample. Data collection was done in duplicate. A 75 μL oil aliquot was deposited onto the heated crystal. Spectra were collected over a range of 4000–700 cm^{-1} at 4 cm^{-1} resolution and by co-adding 64 scans, to improve the signal-to-noise ratio. Spectral data were displayed in terms of absorbance and viewed using Resolutions Pro Software (Agilent, Santa Clara, CA,

USA). Raman Spectroscopy: Olive oil samples were heated (65 °C) in a lab oven before the analysis. Three milliliters of olive oil sample was placed in a quartz cuvette (Hellma Analytics, Mullheim, Germany) with the 10-mm light path for Raman analysis using a WP 1064 compact benchtop Raman spectrometer (Wasatch Photonics, Durham, NC, USA). The Raman spectroscopy was equipped with an Indium Gallium Arsenide (InGaAs) detector and a laser source operating at 1064 nm. The Raman spectra were collected from 250 to 1850 cm^{-1} with a resolution of 4 cm^{-1} and 3 scans were co-added and averaged to improve the signal-to-noise ratio of the spectrum with an integration time of 3000 ms. Between each sample, the background spectrum was acquired to eliminate environmental variations. Spectral data were displayed in terms of scattered light by the sample and viewed using Enlighten^{TM} software (Wasatch Photonics, Durham, NC, USA). Spectral data collection was done in duplicate.

2.4. Multivariate Data Analysis

The spectral data were imported as GRAMS (.spc) and Excel (.xls) files and analyzed using Pirouette$^®$ multivariate statistical analysis software (version 4.5, Infometrix Inc., Bothell, WA, USA). FT-IR spectral data were transformed by smoothing (35 points) and taking the Savitsky–Golay second derivative (35 points with second order polynomial filter). Raman spectral data were preprocessed using mean-center and transformed taking the Savitsky–Golay second derivative (35 points with second order polynomial filter). Samples with high residual and leverage were re-evaluated and excluded if needed. The remaining samples were randomly divided into two sub-groups as calibration (80% of the total sample size) and validation (remaining 20%) sets.

Classification analyses of olive oils were performed by using soft independent modeling of class analogy (SIMCA), a supervised pattern recognition classification technique that uses previous knowledge about the category membership of samples to classify new unknown samples in one of the known classes based on its pattern of measurements [35]. The optimal number of principal components (PCs) for each class in the training set was determined by cross-validation, thus, lessening the effect of noise-laden PCs in the class model [35]. Class boundaries surrounding each class in the multivariate space represented the mean residual standard deviation of the training samples for a given class based on an F-statistic value set at a 95% specific confident interval. Interclass distances measure class separation in the multivariate space and interclass distances between groups of objects above 3.0 is regarded as significant to identify 2 groups of samples as different classes [36]. Lastly, the prediction of class membership was achieved by comparing the residual variance of an unknown to the average residual variance of the classes in the model using an F-test [37]. SIMCA only assigns unknown samples to the class for which it has the smallest residual, not forcing class assignments if the residual variance of an unknown exceeds the upper limit for every modeled class in the dataset. The sample will not be assigned to a class because it is either an outlier or comes from a class not represented in the model [37].

Partial least squares regression (PLSR) models were developed using infrared and Raman spectra and reference values obtained for fatty acid composition, free fatty acids, peroxide value, pyropheophytins, and total polar compounds. Separate PLSR models were developed for the infrared and Raman systems for each of the compounds of interest. PLSR combines features from principal component analysis (PCA) and multiple regression to solve problems involving high collinearity and to determine a set of dependent variables from a (very) large set of independent variables or predictors [38,39]. The PLSR algorithm extracts a set of orthogonal factors called "latent variables" that explains most of the variance from the X (spectra) and Y (concentration), generating an algorithm that diminishes the potential impact of large, irrelevant variations in the X matrix [39]. Leave-one-out cross-validation was applied to determine the optimal number of factors to prevent over- or under-fitting and to improve the modeling performance and the quality of the prediction [38]. The quality of the final model was evaluated based on the number of latent variables, loading vectors, standard error of cross-validation (SECV), the coefficient of determination (R-value), standard error of prediction (SEP), and outlier diagnostics, while outliers were determined using residual and Mahalanobis distances.

The performances of models were determined by calculating the specificity and sensitivity based on true positive (TP, predicted result and actual label are both positive), false positive (FP, predicted result is positive while the actual label is negative), true negative (TN, predicted result and the actual label are both negative) and false negative (FN, predicted result is negative while the actual label is positive) classifiers [40].

3. Results and Discussion

3.1. Characterization of Olive Oils Using International Olive Oil Trade Standards

Olive oils were grouped as extra virgin olive oil (EVOO) ($n = 77$), virgin olive oil (VOO)/olive oil (OO) ($n = 27$), and adulterated olive oil with vegetable oils (corn, sunflower, soybean, and canola oil) ($n = 47$) according to information provided by the Aydin Commodity Exchange Laboratories (Aydin, Turkey) and California Olive Oil Council. Table 1 summarizes the information on reference analysis with regard to the levels of major fatty acids, free fatty acids (FFA), peroxide value (PV), pyropheophytins (PPP), and total polar compounds (TPC).

Table 1. Reference concentration levels for the compounds measured in olive oil samples.

		EVOO [a]	VOO/OO [b]	Mixture [c]
	Range	9.8–17.4	10.6–18.1	5.3–18.9
Palmitic (%)	Mean	13.2	13.4	12.1
	SD	1.7	1.9	2.8
	Range	2.7–2.9	2.7–3.1	2.7–3.5
Stearic (%)	Mean	2.8	2.8	2.9
	SD	0	0.1	0.2
	Range	62.0–78.2	57.7–76.5	11.0–76.9
Oleic (%)	Mean	72.6	71.5	66.9
	SD	3.8	4.4	14
	Range	4.5–14.8	6.0–17.7	5.6–76.0
Linoleic (%)	Mean	8.5	9.5	15.1
	SD	2.2	2.4	14
	Range	0.6–0.8	0.7–0.9	0.1–5.8
Linolenic (%)	Mean	0.7	0.7	1
	SD	0	0.1	0.9
	Range	0.1–0.7	0.1–1.9	0.1–10.3
Free Fatty Acid (%)	Mean	0.4	0.5	2.1
	SD	0.2	0.5	2.7
Peroxide Value	Range	4.8–13.7	3.1–13.2	2.5–32.7
(meqO$_2$/kg)	Mean	9.8	10	11.7
	SD	2	2.5	4.9
Pyropheophytin	Range	7.0–14.9	5.6–20.6	12.5–25.5
(%)	Mean	11.5	13.2	19.8
	SD	2.3	3	3
Total Polar	Range	2.5–8.5	4.0–9.8	5.5–17.8
Compound (%)	Mean	5.2	6.6	8.7
	SD	1.1	1.5	2.4

[a] EVOO: Extra virgin olive oil, [b] VOO/OO: Blend of virgin olive oil and olive oil, [c] Mixture: Adulterated olive oil with vegetable oils (corn, sunflower, soybean, and canola oil).

Fatty acid (FA) composition of the EVOO group (Table 1) showed that the five major FAs (16:0, 18:0, 18:1*n*-9, 18:2*n*-6 and 18:3*n*-3) fell within specified ranges set by the United States standards for grades of olive oil [41] and International Olive Council [42]. EVOO variation in FA levels among samples can be related to differences in geographic origin, variety, stage of maturity of the fruit, latitude, climatic conditions, storage, and extraction process of samples [43–45]. EVOO and VOO showed similar fatty acid profiles, except for a sample obtained from Peru that showed higher palmitic (18.1%) and linoleic (17.7%) but lower oleic (57.7%) compared to other VOO samples. On the contrary, adulterated olive oils with vegetable oils showed marked variation in FA composition (Table 1). For instance, olive oil adulterated with canola oil had lower palmitic acid (5.7%), while linoleic (28.5%) and linolenic (4.4%) acids were higher than pure olive oil. Adulteration of EVOO with corn oil resulted in a decrease in the levels of oleic acid (29.9%) and an increase in linoleic acid (58.6%) content.

The average FFA content of the EVOO and VOO/OO samples ranged from 0.4 ± 0.2% and 0.5 ± 0.5%, respectively. The main difference between EVOO and VOO resulted from their FFA content. According to the trade standards of the International Olive Council (IOC) (2018), the FFA content of EVOO, VOO, and OO cannot exceed 0.8%, 2.0%, and 1.0%, respectively [42]. FFA levels of adulterated EVOO samples with other vegetable oils ranged from 0.1% to 10.3% (2.1 ± 2.7%). In particular, two adulterated EVOO samples showed FFA levels of 9.0% and 10.3% that could be related to mixing olive oils with crude vegetable oil or waste cooking or frying oil. There is no FFA limit for the crude vegetable oils, van Doosselaere (2013) reported that crude palm oil FFA levels could reach levels of 20–25% because of the lipolytic enzymes of the fruit that were not handled properly [46]. The frying or cooking process increases the FFA content of vegetable oils since oils that contain high levels of polyunsaturated fatty acids are highly susceptible to hydrolysis, oxidation, and polymerization under a frying environment [47].

Peroxide value of olive oil samples were 9.8 ± 2.0, and 10.0 ± 2.5 meqO$_2$/kg for EVOO and VOO/OO samples, respectively. According to the European Union Commission Regulations (EEC/2568/91), the PV limit for EVOO and VOO are 20 meqO$_2$/kg, whereas the limit for OO is 15 meqO$_2$/kg [48], and our findings were under the established limits for different grades of olive oils. Similar values for PV of EVOO and VOO, ranging from 6.2 to 11 meqO$_2$/kg, were reported by Casal and others (2010) [49]. A high PV indicates that olives or paste were likely mishandled [50]. Adulterated olive oils with other vegetable oils showed PV ranging from 2.5 to 32.7 meqO$_2$/kg, indicating that counterfeiters employ a wide array of oil quality, including freshly deodorized to highly oxidized vegetable oils.

Pyropheophytin (PPP) values of the samples were 11.5 ± 2.3, 13.2 ± 3.0, 19.8 ± 3.0% for EVOO, VOO/OO blends, and olive oil mixtures with vegetable oil samples, respectively. The PPPs are the breakdown products of chlorophyll in olive oil. The chlorophyll pigment initially breaks down to pheophytin (a and a′), and then into pyropheophytins, due to the decarbomethoxylation of chlorophyll and pheophytins, upon the effect of heat [51]. The elevated level of PPP indicates that the samples were oxidized and/or adulterated with cheaper refined oils and the limit of the total PPP should be lower than 15% in EVOO [52].

Average total polar compounds (TPC) of the EVOO, VOO/OO, and adulterated olive oils ranged from 5.2 ± 1.1%, 6.6 ± 1.5%, and 8.7 ± 2.4%, respectively. The TPC measures the polar fraction in oils that are composed of polymers (dimers, trimers, and highly polymerized compounds) and decomposition products (mono and diacylglycerols, FFAs, volatile compounds, cyclic, and non-cyclic monomers) [53]. The TPC limit for frying oil is 25% according to international legislation, and if an oil exceeds this limit it becomes unsuitable for human consumption [53].

Overall, the chemical quality parameters of EVOO and OO showed strong overlapping within minimum and maximum limits, making it challenging to use these parameters as reliable markers to identify potential adulteration to consumers.

3.2. Spectral Analysis of Olive Oil Samples

The characteristic FT-IR absorption spectra of different grades of olive oil samples and their corresponding band assignments for specific functional groups are displayed in Figure 1a. Visual inspection of the spectra showed close resemblance in their spectral profiles throughout the mid-IR region (4000–700 cm^{-1}) (Figure 1a), similar to those previously reported by Rohman and others (2017) [54]. Key absorbance signals included the band at 3010 cm^{-1} associated with =C–H stretching of cis olefins, the 2900–2800 cm^{-1} range related to–C–H symmetrical and asymmetrical stretching vibrations (CH$_2$ and CH$_3$), the band centered at 1746 cm^{-1} associated to the stretching vibrations of the ester carbonyl (–C=O) functional group of triglycerides, and the band at 1465 cm^{-1} associated with C–H bending (scissoring) vibration of the CH$_2$ group. The band at 1377 cm^{-1} corresponds to the C–H bending (symmetrical) vibration of the CH$_3$ group, and the shoulder band centered at 1417 cm^{-1} due to the rocking vibrations of the C-H bonds of cis-disubstituted olefins. Finally, the fingerprint region from 1200 to 1000 cm^{-1} represented the unique stretching and bending vibrations of –C–O and –CH$_2$– vibrational modes. Overall, important spectral regions for revealing possible EVOO adulteration included the band intensities at 3010–2800 cm^{-1} related to the triglyceride fatty acid composition and level of unsaturation of the oils, and the relative proportion between the triglyceride ester-linkage (COOR) band at 1742 cm^{-1} and the C=O absorption of FFAs at 1711 cm^{-1}. An increase in the band intensity at 1711 cm^{-1} correlates with the increase in FFA content of oil [55].

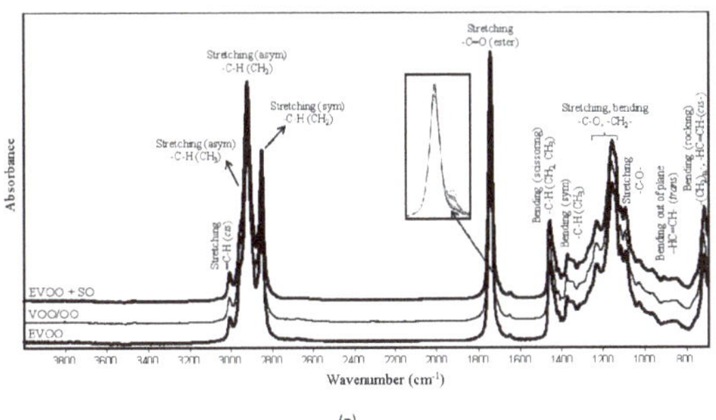

(a)

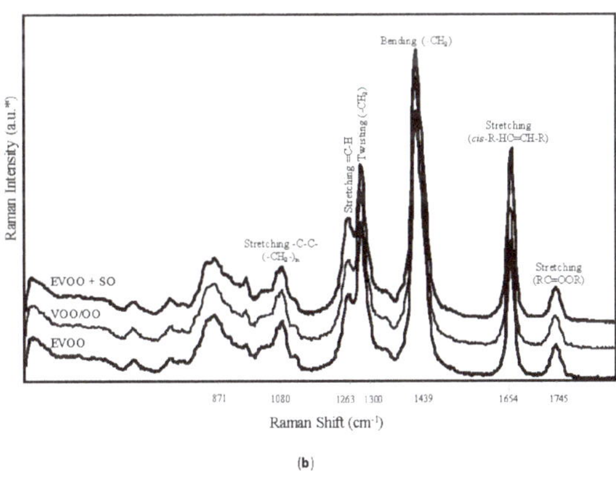

(b)

Figure 1. (a) FT-IR spectrum and band assignments of different quality olive oils at frequency of 4000–700 cm^{-1} collected using a portable 5-reflections ZnSe crystal ATR system equipped with a temperature-controlled accessory. (b) Raman spectrum of different quality olive oils at frequencies of 200–1850 cm^{-1} collected using a compact benchtop Raman system working with 1064 nm excitation laser. EVOO: Extra virgin olive oil, VOO/OO: Blend of virgin olive oil and olive oil, EVOO + SO: Extra virgin olive oil + Sunflower oil. *a.u.: Arbitrary units.

The Raman spectra for selected olive oil samples and their band assignments for specific functional groups are given in Figure 1b. The band at 1080 cm^{-1} was associated with C-C stretching vibration (-CH$_2$-)$_n$, while the band at 1263 cm^{-1} was associated with =C–H in-plane deformation of a conjugated cis double bond (cis-R-HC=CH-R) and related with monounsaturated fatty acids. The band at 1300 cm^{-1} was related to -C-H twisting motion (-CH$_2$), and the band at 1439 cm^{-1} was associated with

-C-H bending (-CH$_2$) modes. The band at 1654 cm^{-1} was related to C=C stretching (cis-R-HC=CH-R) from polyunsaturated fatty acids. The band at 1745 cm^{-1} was associated with C=O stretching vibration (RC=OOR) [9,56]. Different pure olive oils (EVOO, VOO, OO) did not show major differences throughout the measured Raman spectrum (Figure 1b), but olive oil adulterated with other vegetable oils displayed marked differences (higher bands) in the band intensities at 1263 and 1654 cm^{-1}. As mentioned earlier, those bands correspond to monounsaturated and polyunsaturated fatty acids, and an increase in their band intensities has been related to an increasing weight percentage of unsaturated fatty acids in olive oils [9,56].

3.3. Pattern Recognition Modeling Using FT-IR and Raman Spectroscopy

The FT-IR and Raman spectral data were analyzed using soft independent modeling of class analogy (SIMCA) for the authentication of EVOO and detection of adulteration, either by blending with other vegetable oils or replacing of EVOO with lower olive oil grades, such as refined, pomace, or lampante olive oils. Single-class and multi-class pattern recognition strategies were assessed either by using a binary (authentic EVOO vs. VOO/OO blends and EVOO adulterated with vegetable oils) or multiple (authentic EVOO, VOO/OO blends and EVOO adulterated with vegetable oils) class approach based on the information provided by the Aydin Commodity Exchange Laboratories and California Olive Oil Council, along with our reference tests' results.

A multi-class approach was implemented for the FT-IR spectral data that comprised three different groups including EVOO, VOO/OO blends, and adulterated olive oil with vegetable oils. The class projection plot (Figure 2a) showed compact clusters for the EVOO and VOO/OO blends, indicating similar chemical composition among samples in their class, while the marked compositional differences in EVOO adulterated with different vegetable oils were reflected by the large spread of samples in the class projection plot. A SIMCA parameter that correlated to the chemical differences between classes was the interclass distances (ICD) and gave values ranging from 2.6 (EVOO & VOO/OO blends) to 6.1 (VOO/OO blends & EVOO with other vegetable oils) (Table 2). In the SIMCA models, two different classes with an ICD >3 are considered significantly different from each other [36]. Overall, all classes were largely independent of one another, requiring three to five PCs to explain 99% of the variance within groups and the cross-validation showed zero misclassifications, which indicates that the model should be robust and minimizes over-fitting. The SIMCA discriminating power plot (Figure 2c) showed that the clustering of different olive oil grades and adulteration were explained by the bands centered at 2920 and 2850 cm^{-1}, corresponding to CH$_2$ asymmetric and symmetric stretching vibrations, and 1742, 1711, and 1098 cm^{-1},which correspond to the stretching vibrations of the carbonyl bonds (–C=O) in acylglycerides, and the 1670 cm^{-1} band, related to the olefinic trans C=C stretching vibrations.

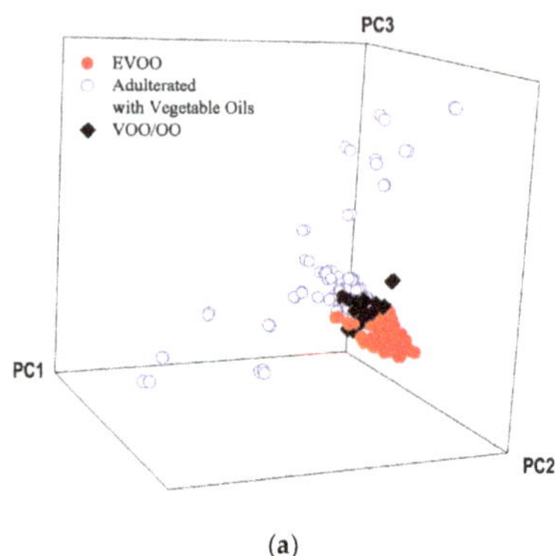

(a)

Figure 2. *Cont.*

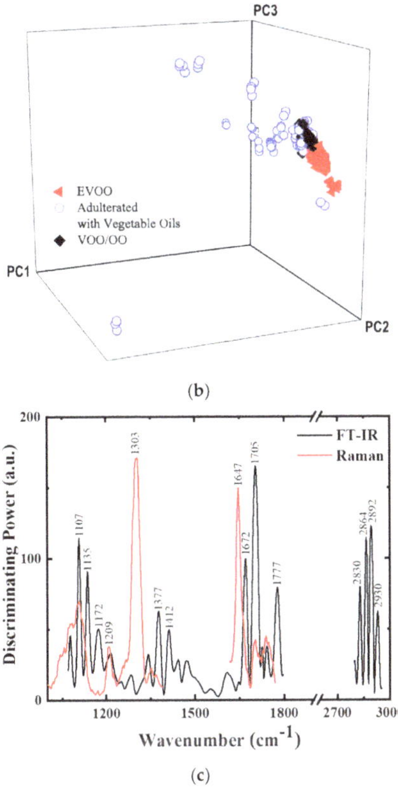

(b)

(c)

Figure 2. (**a**) Soft independent modeling of class analogy (SIMCA) 3D projection plots of spectral data for olive oil samples collected by (**a**) portable FT-IR and (**b**) compact benchtop Raman spectrometers. EVOO: Extra virgin olive oil, VOO/OO: Blend of virgin olive oil and olive oil. (**c**) SIMCA discriminating plot based on the mid-infrared and Raman spectra of olive oils using an FT-IR and a Raman spectrometer, showing bands and regions responsible for class separation.

Table 2. Interclass distances between three classes of olive oils based on the SIMCA class projections for the FT-IR spectra collected in the 700–4000 cm^{-1} region.

Groups	EVOO [a]	VOO/OO Blends [b]	EVOO with other Vegetable Oils [c]
EVOO	0		
VOO/OO blends	2.6	0	
EVOO with other vegetable oils	5.2	6.1	0

[a] EVOO: Extra virgin olive oil, [b] VOO/OO: Blend of virgin olive oil and olive oil, [c] Adulterated EVOO with other vegetable oils (corn, sunflower, soybean, and canola oil).

The predictive performance of the multi-class calibration model was determined by using an independent validation set that included fifteen EVOOs, five VOO/OO blends, and nine EVOOs adulterated with other vegetable oils. By including the information of additional classes (i.e., VOO/OO blends and EVOO with other vegetable oils), the sensitivity and specificity of the SIMCA models were 100% for all the oil classes (Table 3). Since authentication studies are often approached as a one-class classification analysis, the adulterants are usually unknown [57]. A one-class SIMCA model was developed for EVOO based on the infrared spectra of genuine samples, and any adulterated samples were classified as outliers when tested against the PCA model boundaries. The performance of the calibration models was evaluated by using an independent validation set that consisted of 15 authentic EVOO and 74 non-authentic (VOO/OO and EVOO with other vegetable oils) samples. All EVOO samples were correctly predicted (TP = 15 and FN = 0) as belonging to its target class, resulting in 100% sensitivity, indicating that the one-class model was capable of accurately identifying authentic EVOO samples. On the other hand, eight of the non-authentic samples were predicted as EVOO (FP = 8, TN = 66), resulting in 89% specificity (Table 3), revealing that the model had adequate ability to detect adulterated samples. The one-class model correctly predicted all EVOO mixed with cheaper vegetable oils, while eight out of twenty-seven VOO/OO were predicted as belonging to the EVOO class.

Table 3. Sensitivity and specificity values of SIMCA multi- and single-class models obtained from FT-IR and Raman spectroscopy.

Model Types		Samples	Sensitivity (%)	Specificity (%)
Multi-Class	FT-IR	VOO/OO blends [b]	100	100
		EVOO [a] with other vegetable oils	100	100
	Raman	VOO/OO blends	100	100
		EVOO with other vegetable oils [c]	100	100
One-Class		FT-IR	100	89
		Raman	100	66

[a] EVOO: Extra virgin olive oil, [b] VOO/OO: Blend of virgin olive oil and olive oil, [c] Adulterated EVOO with other vegetable oils (corn, sunflower, soybean, and canola oil).

A similar approach was taken for the Raman spectral data collected from the oils to detect EVOO adulteration. The class projection plot is given in Figure 2b. The multi-class SIMCA model gave ICDs ranging from 0.9 to 7.0, with the largest dissimilarity of spectral features obtained between authentic EVOO and its mixtures with other vegetable oils (ICD = 7.0), while the ICD differentiating EVOO from VOO and its blends with refined olive oils was 0.9 (Table 4). Wold and Sjöström (1977) described that distances between class models larger than one indicate real differences, and if two models are not independent, the interclass distance is close to zero [58]. The classes required three to five PCs to explain 98% of the variance within groups, and the cross-validation showed zero misclassifications. The SIMCA discriminating power plot (Figure 2c) was dominated by the bands centered at 1652 and 1306 cm^{-1}, associated with the alkene νC=C stretch and in-phase methylene twisting vibrations, respectively. The minor bands at 920 and 856 cm^{-1} were attributed with bending vibrations of trans (C=C) and stretching vibrations of methylene chain skeleton, respectively [8]. An independent validation set was used to evaluate the predictive performance of the SIMCA models. Sensitivity evaluated the capability of our classification model to identify EVOO, while specificity determined the ability of our model to discriminate the adulterated or mislabeled samples. The sensitivity and specificity values for the single and multi-class models for Raman spectroscopy are given in Table 3. The multi-class model gave 100% sensitivity and specificity, which means that models generated by Raman spectra could effectively detect authentic EVOO samples from adulterated oils with excellent accuracy. Although the ICD separating the pure EVOO from VOO and its blends with refined olive oils was 0.9, the model gave perfect predictions. SIMCA single class models developed from Raman models correctly predicted all authentic EVOO (TP = 15 and FN = 0; 100% sensitivity). However, out of the 74 validation samples that were either mislabeled (lower olive oil grades) or adulterated with other vegetable oils, the one-class model failed to identify 25 samples that were predicted as pure EVOO (FP = 25, TN = 49; sensitivity = 66%). A total of 12 VOO/OO blends and 13 adulterated samples were classified as EVOO.

Table 4. Interclass distances between three classes of olive oils based on the SIMCA class projections for the Raman spectra collected in the 250–1850 cm^{-1} region.

Groups	EVOO [a]	VOO/OO Blends [b]	EVOO with other Vegetable Oils [c]
EVOO	0		
VOO/OO blends	0.9	0	
EVOO with other vegetable oils	7.0	5.9	0

[a] EVOO: Extra virgin olive oil, [b] VOO/OO: Blend of virgin olive oil and olive oil, [c] Adulterated EVOO with other vegetable oils (corn, sunflower, soybean, and canola oil).

Similar to our findings, Li et al. (2018), Philippidis et al. (2017), and Zhang et al. (2011) were also be able to differentiate olive oils from vegetable oils including waste cooking oil, sunflower, rapeseed, soybean, corn, and canola oil by using Raman spectroscopy [8,9,56]. However, we report for the first time the discrimination of EVOO from their different grades (VOO and OO). Our data showed the challenges in detecting EVOO from OO, as very few unique compounds, monochloropropanediol esters, and glycidyl esters formed in the refining process can be used as markers for authentication [59]. By including the additional features from the class assigned to VOO and OO samples to the supervised model allowed to improve the discriminability of the classifiers providing the best accuracy for authentication of EVOO without false positives. Furthermore, EVOO adulterated with pomace olive oil showed marked FT-IR and Raman spectral differences allowing straightforward detection by pattern recognition analysis.

3.4. Development of PLSR Models Using FT-IR and Raman Spectroscopy

Extra virgin olive oil (EVOO) quality and its freshness degrade over time due to its high level of monounsaturated fatty acid content (oleic acid). Therefore, it is important to monitor the main quality parameters (FFA, PV, PPP, TPC, and major fatty acid content) in EVOO throughout the olive oil production process and during the storage. Taking this into account, the FT-IR and the Raman spectra collected using the portable and compact benchtop units were employed to develop quantitative models with partial least squares regression (PLSR) based on reference values for free fatty acids (FFA), peroxide value (PV), pyropheophytin (PPP), total polar compounds (TPC), and major fatty acids (palmitic, stearic, oleic, linoleic, and linolenic) (Figure 3). Samples were randomly divided into two groups as calibration and external validation sets, eighty percent of the total number of samples were randomly chosen to generate the calibration set and the other twenty percent were used to generate the external validation set to assess the robustness of the models. The performance statistics of each model, the minimum and maximum values, and the number of samples used in each calibration and external validation set were given in Table 5. If a sample has high leverage and/or residual, it was identified as an outlier and excluded from the model, therefore the total number of samples in each model could be different from each other. For the best model performances, and to eliminate the irrelevant, noisy, and unreliable variables (wavenumbers), specific wavenumbers were selected from the FT-IR and Raman spectral regions for each analyte. Depending on the quality parameter, cross-validation (leave-one-out) identified three to six factors to generate the FT-IR and Raman calibration models.

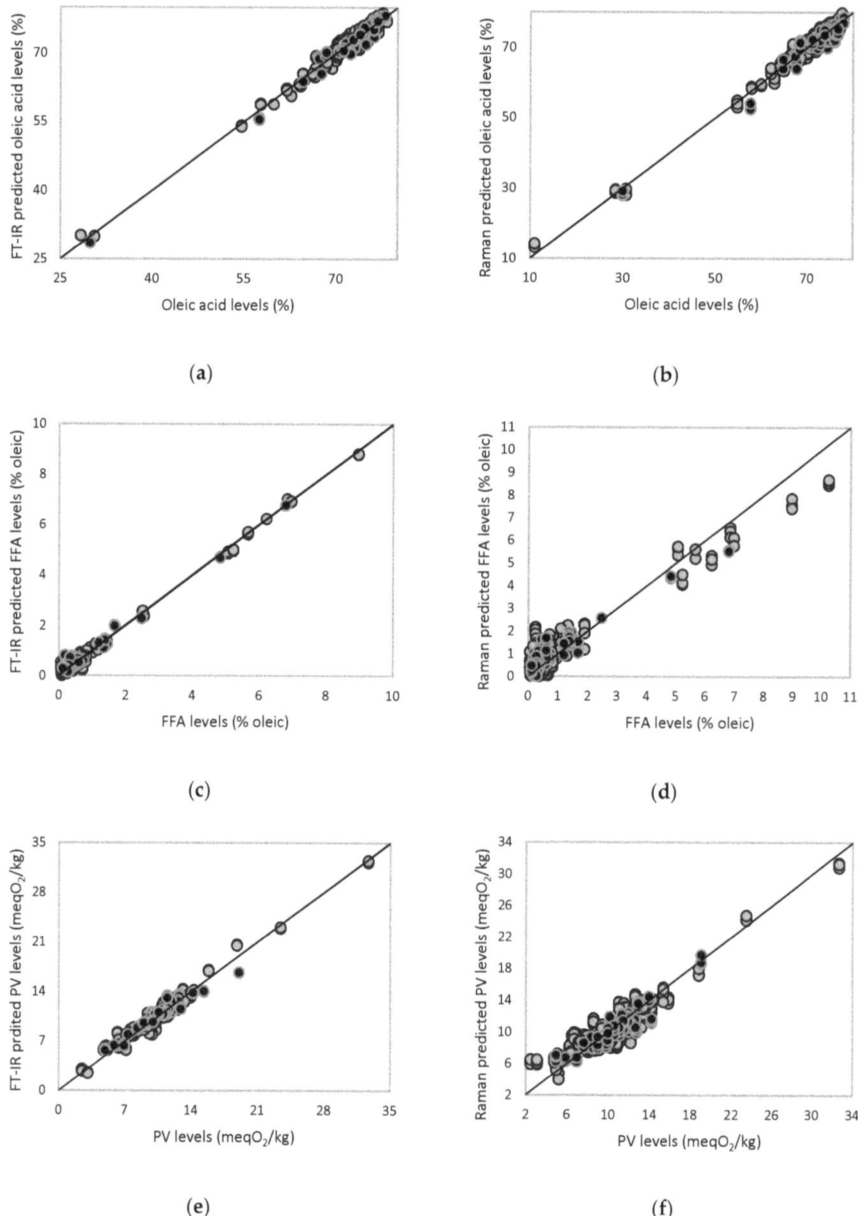

Figure 3. Partial least squares regression (PLSR) calibration and external validation plots for oleic (**a** and **b**), free fatty acids (**c** and **d**), and peroxide value (**e** and **f**) levels in olive oil samples using a portable 5-reflections FT-IR and compact benchtop Raman instrument, respectively. Grey circles represent samples in calibration set; black circles represent samples in external validation set.

Table 5. Performance statistics of calibration and external validation models developed by using portable FT-IR and compact benchtop Raman spectroscopy.

Technique	Parameter	Calibration Model					External Validation Model			
		Range	N [a]	Factor	SECV [b]	Rcal	Range	N [c]	SEP [d]	Rval
FT-IR	Palmitic (%)	5.3–18.9	120	6	0.44	0.98	6.5–18.1	30	0.53	0.98
	Stearic (%)	2.7–3.6	120	4	0.03	0.98	2.7–3.5	30	0.02	0.99
	Oleic (%)	11.0–78.2	120	4	1.13	0.99	29.9–78.0	30	1.41	0.99
	Linoleic (%)	4.5–76.0	120	4	1	0.99	5.7–41.0	30	1.4	0.98
	Linolenic (%)	0.5–1.8	117	4	0.02	0.99	0.6–1.0	29	0.02	0.97
	FFA (%)	0.1–10.3	118	3	0.17	1	0.1–6.8	30	0.23	0.99
	PV (meqO$_2$/kg)	2.5–32.7	120	5	0.65	0.98	4.9–19.1	30	0.79	0.96
	Pyropheophytin (%)	5.6–25.5	87	6	1.47	0.96	10.7–23.5	22	1.46	0.94
	TPC (%)	2.5–17.8	120	6	0.54	0.97	3.3–13.3	30	0.59	0.97
Raman	Palmitic (%)	5.3–18.9	120	6	0.84	0.91	6.5–18.1	30	0.99	0.92
	Stearic (%)	2.7–3.6	120	5	0.04	0.96	2.7–3.5	30	0.04	0.97
	Oleic (%)	11.0–78.2	120	6	1.33	0.99	29.9–78.0	30	1.78	0.98
	Linoleic (%)	4.5–76.0	120	4	1.09	0.99	5.7–41.0	30	1.63	0.99
	Linolenic (%)	0.5–1.8	118	6	0.02	0.99	0.6–1.0	30	0.01	0.98
	FFA (%)	0.1–10.3	118	6	0.55	0.94	0.1–6.8	30	0.52	0.93
	PV (meqO$_2$/kg)	2.5–32.7	120	4	1.31	0.92	4.9–19.1	30	1.11	0.92
	Pyropheophytin (%)	7.0–25.5	85	5	1.93	0.92	10.7–20.5	21	1.55	0.92
	TPC (%)	2.5–17.8	119	6	0.76	0.94	3.3–13.3	30	0.83	0.93

[a] Number of samples used in calibration models. [b] Standard error of cross validation. [c] Number of samples used in external validation models. [d] Standard error of prediction.

Table 5 shows the performance statistics for the PLSR calibration and external validation models that were obtained for five major fatty acids (palmitic, stearic, oleic, linoleic, and linolenic) tested in olive oils and the main indices (FFA, PV, PPP, and TPC) that monitor olive oil quality. The SECV values for each calibration model was similar to the standard error of prediction (SEP) of their corresponding external validation model (Table 5), demonstrating the robustness of the generated models. The SEP values ranged from 0.01% to 1.5% for the five major fatty acids present in the tested olive oils. Our models showed superior performance statistics for the estimation of fatty acid profiles (lower correlation coefficient and SEP) than those reported by Gurdeniz and others (2010) for extra virgin olive oils using a benchtop FT-IR unit [60]. Furthermore, our calibration and validation models for the major fatty acids had similar performances to those reported by [61], but they employed 13–14 factors to acquire those statistics, which probably over-fitted the models. Using the same FT-IR and Raman spectral data, we also generated models for the main olive oil quality indices including FFA, PV, PPP and TPC and their performance statistics are given in Table 5. Overall, the FT-IR regression models gave superior performance than those generated by Raman spectroscopy. For example, the model generated by FT-IR for estimation of FFA levels gave correlation coefficient of validation (R_v) of 1.00 and standard error of prediction (SEP) of 0.23% by using three factors, while the Raman model gave an R_v of 0.93 and SEP of 0.55 by using six factors (Table 5). Gouvinhas and others (2015) obtained good performances ($R^2 = 0.99$) on the prediction of FFA content in EVOO at different maturation stages by using a shorter excitation wavelength laser (488 nm) over the spectral range of 950–1800 cm^{-1} [62].

4. Conclusions

The present study was designed to evaluate portable FT-IR and compact benchtop Raman technology for the nondestructive authentication of premium EVOO and detect adulteration with the addition of lower grades of olive oils or other vegetable oils. Multi–class pattern recognition algorithms defining EVOO, VOO/OO (lower quality olive oils), and adulterated EVOO with vegetable oils classes allowed accurate classification with perfect sensitivity and specificity. However, a single-class approach resulted in diminished sensitivity, resulting in the misclassification of VOO and OO samples as EVOO. Our data demonstrated the importance of developing supervised classification models, including relevant a priori knowledge in the training set, especially samples with similar compositional make-up, such as lower quality olive oils, to develop reliable methods to reveal EVOO fraud. Furthermore, the same spectra were used to generate multivariate regression models to predict major quality parameters, including levels of fatty acids, %FFA, PV, PPP, and TPC. Both the portable FT-IR and compact benchtop 1064 nm Raman were promising technologies for "in-situ", non-destructive, simple and quick identification of possible adulteration of EVOOs. However, the portable FT-IR unit gave the best classification and quantitation results, even when comparing against reported SEP collected in benchtop systems. Our approach showed sensitivity and specificity to detect EVOO fraud, even with lower processing grade olive oils, and provides rapid quantitative analysis for monitoring oil quality parameters.

Author Contributions: D.P.A.; Formal Analysis, Methodology, Data curation, Original draft, A.D.K.; Methodology, Resources, B.K.; Resources, L.R.-S.; Data curation, Review and editing. All authors have read and agreed to the published version of the manuscript.

Funding: This research received no external funding.

Acknowledgments: We would like to thank the Turkish Ministry of Education for supporting this project through their PhD fellowship. The authors also are grateful to Aydin Commodity Exchange Laboratories (Aydin, Turkey) and the California Olive Oil Council for providing samples.

Conflicts of Interest: Authors declare that they have no conflict of interest. This article does not contain any studies with human or animal subjects.

References

1. Gelpí, E.; Posada de la Paz, M.; Terracini, B.; Abaitua, I.; Gómez de la Cámara, A.; Kilbourne, E.M.; Lahoz, C.; Nemery, B.; Philen, R.M.; Soldevilla, L.; et al. The spanish toxic oil syndrome 20 years after its onset: A multidisciplinary review of scientific knowledge. *Environ. Health Perspect.* **2002**, *110*, 457–464. [CrossRef]

2. Bajoub, A.; Bendini, A.; Fernández–Gutiérrez, A.; Carrasco–Pancorbo, A. Olive oil authentication: A comparative analysis of regulatory frameworks with especial emphasis on quality and authenticity indices, and recent analytical techniques developed for their assessment. A. review. *Crit. Rev. Food Sci. Nutr.* **2018**, *58*, 832–857. [CrossRef]

3. United States International Trade Commission. *Olive Oil: Conditions of Competition between U.S. and Major Foreign Supplier Industries*; United States International Trade Commission: Washington, DC, USA, 2013.

4. Aparicio, R.; Morales, M.T.; Aparicio–Ruiz, R.; Tena, N.; García–González, D.L. Authenticity of olive oil: Mapping and comparing official methods and promising alternatives. *Food Res. Int.* **2013**, *54*, 2025–2038. [CrossRef]

5. Milanez, K.D.T.M.; Nóbrega, T.C.A.; Nascimento, D.S.; Insausti, M.; Band, B.S.F.; Pontes, M.J.C. Multivariate modeling for detecting adulteration of extra virgin olive oil with soybean oil using fluorescence and UV–Vis spectroscopies: A preliminary approach. *LWT-Food Sci. Technol.* **2017**, *85*, 9–15. [CrossRef]

6. Aroca–Santos, R.; Cancilla, J.C.; Pérez–Pérez, A.; Moral, A.; Torrecilla, J.S. Quantifying binary and ternary mixtures of monovarietal extra virgin olive oils with UV–vis absorption and chemometrics. *Sens. Actuators B Chem.* **2016**, *234*, 115–121. [CrossRef]

7. Durán Merás, I.; Domínguez Manzano, J.; Airado Rodríguez, D.; Muñoz de la Peña, A. Detection and quantification of extra virgin olive oil adulteration by means of autofluorescence excitation–emission profiles combined with multi–way classification. *Talanta* **2018**, *178*, 751–762. [CrossRef] [PubMed]

8. Li, Y.; Fang, T.; Zhu, S.; Huang, F.; Chen, Z.; Wang, Y. Detection of olive oil adulteration with waste cooking oil via Raman spectroscopy combined with iPLS and SiPLS, Spectrochim. *Acta-Part A Mol. Biomol. Spectrosc.* **2018**, *189*, 37–43. [CrossRef] [PubMed]

9. Philippidis, A.; Poulakis, E.; Papadaki, A.; Velegrakis, M. Comparative Study using Raman and Visible Spectroscopy of Cretan Extra Virgin Olive Oil Adulteration with Sunflower Oil. *Anal. Lett.* **2017**, *50*, 1182–1195. [CrossRef]

10. Mendes, T.O.; da Rocha, R.A.; Porto, B.L.S.; de Oliveira, M.A.L.; dos Anjos, V.D.C.; Bell, M.J. Quantification of Extra–virgin Olive Oil Adulteration with Soybean Oil: A Comparative Study of NIR, MIR, and Raman Spectroscopy Associated with Chemometric Approaches. *Food Anal. Methods* **2015**, *8*, 2339–2346. [CrossRef]

11. Rohman, A.; Man, Y.B.C. Fourier transform infrared (FTIR) spectroscopy for analysis of extra virgin olive oil adulterated with palm oil. *Food Res. Int.* **2010**, *43*, 886–892. [CrossRef]

12. Di Girolamo, F.; Masotti, A.; Lante, I.; Scapaticci, M.; Calvano, C.D.; Zambonin, C.; Muraca, M.; Putignani, L. A simple and effective mass spectrometric approach to identify the adulteration of the mediterranean diet component extra–virgin olive oil with corn oil. *Int. J. Mol. Sci.* **2015**, *16*, 20896–20912. [CrossRef] [PubMed]

13. Calvano, C.D.; De Ceglie, C.; D'Accolti, L.; Zambonin, C.G. MALDI–TOF mass spectrometry detection of extra–virgin olive oil adulteration with hazelnut oil by analysis of phospholipids using an ionic liquid as matrix and extraction solvent. *Food Chem.* **2012**, *134*, 1192–1198. [CrossRef] [PubMed]

14. Lorenzo, I.M.; Pavón, J.L.P.; Laespada, M.E.F.; Pinto, C.G.; Cordero, B.M. Detection of adulterants in olive oil by headspace–mass spectrometry. *J. Chromatogr. A* **2002**, *945*, 221–230. [CrossRef]

15. Šmejkalová, D.; Piccolo, A. High–power gradient diffusion NMR spectroscopy for the rapid assessment of extra–virgin olive oil adulteration. *Food Chem.* **2010**, *118*, 153–158. [CrossRef]

16. Xu, Z.; Morris, R.H.; Bencsik, M.; Newton, M.I. Detection of virgin olive oil adulteration using low field unilateral NMR. *Sensors* **2014**, *14*, 2028–2035. [CrossRef]

17. Vigli, G.; Philippidis, A.; Spyros, A.; Dais, P. Classification of edible oils by employing 31P and 1H NMR spectroscopy in combination with multivariate statistical analysis. A proposal for the detection of seed oil adulteration in virgin olive oils. *J. Agric. Food Chem.* **2003**, *51*, 5715–5722. [CrossRef]

18. Fragaki, G.; Spyros, A.; Siragakis, G.; Salivaras, E.; Dais, P. Detection of extra virgin olive oil adulteration with lampante olive oil and refined olive oil using nuclear magnetic resonance spectroscopy and multivariate statistical analysis. *J. Agric. Food Chem.* **2005**, *53*, 2810–2816. [CrossRef]

19. Dugo, G.; Rotondo, A.; Mallamace, D.; Cicero, N.; Salvo, A.; Rotondo, E.; Corsaro, C.; Mannina, L.; Salvo, A. Enhanced detection of aldehydes in extra–virgin olive oil by means of band selective NMR spectroscopy. *Physica A* **2015**, *420*, 258–264. [CrossRef]

20. Rotondo, A.; Mannina, L.; Salvo, A. Multiple Assignment Recovered Analysis (MARA) NMR for a Direct Food Labeling: The Case Study of Olive Oils. *Food Anal. Methods.* **2019**, *12*, 1238–1245. [CrossRef]

21. Kumar, S.; Kahlon, T.; Chaudhary, S. A rapid screening for adulterants in olive oil using DNA barcodes. *Food Chem.* **2011**, *127*, 1335–1341. [CrossRef]

22. Mildner–Szkudlarz, S.; Jeleń, H.H. Detection of olive oil adulteration with rapeseed and sunflower oils using mos electronic nose and smpe–ms. *J. Food Qual.* **2010**, *33*, 21–41. [CrossRef]

23. Knolhoff, A.M.; Croley, T.R. Non–targeted screening approaches for contaminants and adulterants in food using liquid chromatography hyphenated to high resolution mass spectrometry. *J. Chromatogr. A* **2016**, *1428*, 86–96. [CrossRef] [PubMed]

24. Tengstrand, E.; Rosén, J.; Hellenäs, K.-E.; Åberg, K.M. A concept study on non–targeted screening for chemical contaminants in food using liquid chromatography–mass spectrometry in combination with a metabolomics approach. *Anal. Bioanal. Chem.* **2013**, *405*, 1237–1243. [CrossRef] [PubMed]

25. Coates, J. A Review of New Small–Scale Technologies for Near Infrared Measurements. Available online: https://www.americanpharmaceuticalreview.com/Featured-Articles/163573-A-Review-of-New-Small-Scale-Technologies-for-Near-Infrared-Measurements/ (accessed on 10 July 2019).

26. Ellis, D.I.; Muhamadali, H.; Haughey, S.A.; Elliott, C.T.; Goodacre, R. Point–and–shoot: Rapid quantitative detection methods for on–site food fraud analysis–moving out of the laboratory and into the food supply chain. *Anal. Methods.* **2015**, *7*, 9401–9414. [CrossRef]

27. Mossoba, M.M.; Azizian, H.; Fardin–Kia, A.R.; Karunathilaka, S.R.; Kramer, J.K.G. First Application of Newly Developed FT–NIR Spectroscopic Methodology to Predict Authenticity of Extra Virgin Olive Oil Retail Products in the USA. *Lipids* **2017**, *52*, 443–455. [CrossRef]

28. Hirri, A.; Gammouh, M.; Gorfti, A.; Kzaiber, F.; Bassbasi, M.; Souhassou, S.; Balouki, A.; Oussama, A. The use of Fourier transform mid infrared (FT–MIR) spectroscopy for detection and estimation of extra virgin olive oil adulteration with old olive oil. *Sky J. Food Sci.* **2015**, *4*, 60–66.

29. Pan, M.; Sun, S.; Zhou, Q.; Chen, J. A Simple and Portable Screening Method for Adulterated Olive Oils Using the Hand–Held FTIR Spectrometer and Chemometrics Tools. *J. Food Sci.* **2018**, *83*, 1605–1612. [CrossRef]

30. Jiménez-Carvelo, A.M.; Osorio, M.T.; Koidis, A.; González-Casado, A.; Cuadros-Rodríguez, L. Chemometric classification and quantification of olive oil in blends with any edible vegetable oils using FTIR–ATR and Raman spectroscopy. *LWT-Food Sci. Technol.* **2017**, *86*, 174–184. [CrossRef]

31. Georgouli, K.; Martinez Del Rincon, J.; Koidis, A. Continuous statistical modelling for rapid detection of adulteration of extra virgin olive oil using mid infrared and Raman spectroscopic data. *Food Chem.* **2017**, *217*, 735–742. [CrossRef]

32. Method Cd 8-53: Peroxide Value-Acetic Acid-Chloroform Method. In *Official Methods and Recommended Practices of the American Oil Chemists' Society*, 6th ed.; Firestone, D. (Ed.) AOCS Press: Urbana, IL, USA, 2012.

33. European Pharmacopoeia, European Pharmacopoeia 01/2005:20501, in: Counc. Eur., vol. 1. 2004. Available online: https://kampoeng2013.files.wordpress.com/2015/12/european-pharmacopoeia-5-with-all-supplements.pdf (accessed on 17 December 2019).

34. ISO 29841:2009/AMD 1:2016, Vegetable Fats and Oils—Determination of the Degradation Products of Chlorophylls a and a' (Pheophytins a, a' and Pyropheophytins)—Amendment 1. 2016. Available online: https://www.iso.org/standard/66396.html (accessed on 12 December 2019).

35. Berrueta, L.A.; Alonso–Salces, R.M.; Héberger, K. Supervised pattern recognition in food analysis. *J. Chromatogr. A* **2007**, *1158*, 196–214. [CrossRef]

36. Kvalheim, O.M.; Karstand, T.V. SIMCA—Classification by Means of Disjoint Cross Validated Principal Components Models. In *Multivariate Pattern Recognition in Chemometrics: Illustrated by Case Studies*; Brereton, R.G., Ed.; Elsevier: Amsterdam, The Netherlands, 1992; pp. 209–248.

37. Checa–Moreno, R.; Manzano, E.; Capitán–Vallvey, L.F. Characterisation and classification of binders used in art materials at the class and the subclass level. *Anal. Methods.* **2014**, *6*, 4009–4021. [CrossRef]

38. Abdi, H. Partial least squares regression and projection on latent structure regression (PLS Regression). *Wiley Interdiscip. Rev. Comput. Stat.* **2010**, *2*, 97–106. [CrossRef]

39. Romía, M.B.; Bernàrdez, M.A. Multivariate Calibration for Quantitative Analysis. In *Infrared Spectroscopy Food Qulity Analytical Control*, 1st ed.; Sun, D.-W., Ed.; Elsevier: New York, NY, USA, 2009; pp. 51–82.

40. Feng, S.; He, A.; Wang, D.; Kang, B. Diagnostic significance of miR–210 as a potential tumor biomarker of human cancer detection: An updated pooled analysis of 30 articles. *Onco Targets Ther.* **2019**, *12*, 479–493. [CrossRef] [PubMed]

41. USDA. United States Standards for Grades of Olive Oil and Olive–Pomace Oil; 2010; pp. 9–10. Available online: https://www.ams.usda.gov/sites/default/files/media/Olive_Oil_and_Olive--Pomace_Oil_Standard%5B1%5D.pdf (accessed on 10 January 2019).

42. International Olive Council. *Trade Standards Applying to Olive Oils and Olive Pomace Oils*; International Olive Council: Madrid, Spain, 2006. Available online: https://portal.ct.gov/-/media/DCP/pdf/OliveOilIOCTradeStandardpdf.pdf?la=en (accessed on 7 May 2019).

43. Boudour–Benrachou, N.; Plard, J.; Pinatel, C.; Artaud, J.; Dupuy, N. Fatty Acid Compositions of Olive Oils from Six Cultivars from East and South–Western Algeria. *Adv. Food Technol. Nutr. Sci. Open J.* **2017**, *3*, 1–5. [CrossRef]

44. Cicero, N.; Albergamo, A.; Salvo, A.; Bua, G.D.; Bartolomeo, G.; Mangano, V.; Rotondo, A.; Di Stefano, V.; Di Bella, G.; Dugo, G. Chemical characterization of a variety of cold–pressed gourmet oils available on the Brazilian market. *Food Res. Int.* **2018**, *109*, 517–525. [CrossRef] [PubMed]

45. Klikarová, J.; Rotondo, A.; Cacciola, F.; Češlová, L.; Dugo, P.; Mondello, L.; Rigano, F. The Phenolic Fraction of Italian Extra Virgin Olive Oils: Elucidation Through Combined Liquid Chromatography and NMR Approaches. *Food Anal. Methods.* **2019**, *12*, 1759–1770. [CrossRef]

46. Van Doosselaere, P. Production of Oils. In *Edible Oil Process*, 2nd Ed.; Hamm, W., Hamilton, R.J., Calliauw, G., Eds.; John Wiley & Sons, Ltd.: West Sussex, UK, 2013; p. 92.

47. Choe, E.; Min, D.B. Chemistry of deep–fat frying oils. *J. Food Sci.* **2007**, *72*, R77–R86. [CrossRef]

48. Commission Regulation (EEC) No 2568/91. 1991. Available online: https://eur-lex.europa.eu/legal-content/EN/TXT/PDF/?uri=CELEX:01991R2568-20151016&from=EN (accessed on 5 December 2019).

49. Casal, S.; Malheiro, R.; Sendas, A.; Oliveira, B.P.P.; Pereira, J.A. Olive oil stability under deep–frying conditions. *Food Chem. Toxicol.* **2010**, *48*, 2972–2979. [CrossRef]

50. Vossen, P. Growing Olives for Oil. In *Handbook Olive Oil Analysis Properties*, 2nd ed.; Aparicio, R., Harwood, J., Eds.; Springer: Berlin/Heidelberg, Germany, 2013; pp. 19–56.

51. Guillaume, C.; Gertz, C.; Ravetti, L. Pyropheophytin a and 1,2–diacyl–glycerols over time under different storage conditions in natural olive oils. *JAOCS J. Am. Oil Chem. Soc.* **2014**, *91*, 697–709. [CrossRef]
52. Frankel, E.N.; Mailer, R.J.; Schoemaker, C.F.; Wang, S.C.; Flynn, J.D. Tests Indicate that Imported "Extra Virgin" Olive Oil often Fails International and USDA Standards. 2010. Available online: https://www.kretareserve.com/info/oliveoilappendix071510.pdf (accessed on 10 July 2019).
53. Osawa, C.C.; Gonçalves, L.A.G.; Gumerato, H.F.; Mendes, F.M. Study of the effectiveness of quick tests based on physical properties for the evaluation of used frying oil. *Food Control* **2012**, *26*, 525–530. [CrossRef]
54. Rohman, A.; bin Che Man, Y.; Ismail, A.; Hashim, P. FTIR spectroscopy coupled with chemometrics of multivariate calibration and discriminant analysis for authentication of extra virgin olive oil. *Int. J. Food Prop.* **2017**, *20*, S1173–S1181. [CrossRef]
55. Bendini, A.; Cerretani, L.; Di Virgilio, F.; Belloni, P.; Bonoli–Carbognin, M.; Lercker, G. Preliminary evaluation of the application of the ftir spectroscopy to control the geographic origin and quality of virgin olive oils. *J. Food Qual.* **2007**, *30*, 424–437. [CrossRef]
56. Zhang, X.F.; Zou, M.Q.; Qi, X.H.; Liu, F.; Zhang, C.; Yin, F. Quantitative detection of adulterated olive oil by Raman spectroscopy and chemometrics. *J. Raman Spectrosc.* **2011**, *42*, 1784–1788. [CrossRef]
57. Zhang, L.; Li, P.; Sun, X.; Mao, J.; Ma, F.; Ding, X.; Zhang, Q. One–class classification based authentication of peanut oils by fatty acid profiles. *RSC Adv.* **2015**, *5*, 85046–85051. [CrossRef]
58. Wold, S.; Sjostrom, M. SIMCA: A method for analyzing chemical data in terms of similarity and analogy. In *Chemometrics Theory and Application, American Chemical Society Symposium Series 52*; Kowalski, B.R., Ed.; American Chemical Society: Washington, DC, USA, 1977; Volume 52, pp. 243–282. [CrossRef]
59. Yan, J.; Oey, S.B.; van Leeuwen, S.P.J.; van Ruth, S.M. Discrimination of processing grades of olive oil and other vegetable oils by monochloropropanediol esters and glycidyl esters. *Food Chem.* **2018**, *248*, 93–100. [CrossRef]
60. Gurdeniz, G.; Ozen, B.; Tokatli, F. Comparison of fatty acid profiles and mid–infrared spectral data for classification of olive oils. *Eur. J. Lipid Sci. Technol.* **2010**, *112*, 218–226. [CrossRef]
61. Maggio, R.M.; Kaufman, T.S.; Del Carlo, M.; Cerretani, L.; Bendini, A.; Cichelli, A.; Compagnone, D. Monitoring of fatty acid composition in virgin olive oil by Fourier transformed infrared spectroscopy coupled with partial least squares. *Food Chem.* **2009**, *114*, 1549–1554. [CrossRef]
62. Gouvinhas, I.; Machado, N.; Carvalho, T.; De Almeida, J.M.M.M.; Barros, A.I.R.N.A. Short wavelength Raman spectroscopy applied to the discrimination and characterization of three cultivars of extra virgin olive oils in different maturation stages. *Talanta* **2015**, *132*, 829–835. [CrossRef]

 *foods* 

Article

Comparison and Discrimination of Two Major Monocultivar Extra Virgin Olive Oils in the Southern Region of Peloponnese, According to Specific Compositional/Traceability Markers

Vasiliki Skiada [1,*], Panagiotis Tsarouhas [2] and Theodoros Varzakas [1,*]

[1] Department of Food Science and Technology, University of Peloponnese, Antikalamos, 24100 Kalamata, Greece
[2] Department of Supply Chain Management (Logistics), International Hellenic University, Kanellopoulou 2, 60100 Katerini, Greece; ptsarouhas@ihu.gr
* Correspondence: vpskiada@yahoo.gr (V.S.); theovarzakas@yahoo.gr (T.V.)

Received: 11 January 2020; Accepted: 4 February 2020; Published: 6 February 2020

Abstract: The qualitative characteristics and chemical parameters were determined for 112 virgin olive oil samples of the two dominant olive cultivars in the southern region of Peloponnese, cv. Koroneiki and cv. Mastoides. As no relevant data exist for this geographical area, yet one of the most important olive-growing regions in Greece, this study aimed to evaluate and evidence the differences on specific chemical characteristics of the oils because of their botanical origin. Olive oils of Koroneiki variety were characterized by a three-fold lower concentration in heptadecanoic and heptadecenoic acid compared to oils of cv. Mastoides. In addition, Mastoides oils exhibited higher β-sitosterol and total sterols concentration and lower Δ-5-avenasterol and total erythodiol content compared to Koroneiki olive oils Analysis of variance and principal component analysis of the GC-analyzed olive oil samples showed substantial compositional differences in the fatty acid and sterolic profile between Koroneiki and Mastoides cultivars. Hence, results demonstrate that the fatty acid and sterolic profile can be used as exceptional compositional marker for olive oil authenticity.

Keywords: EVOO; cv. Koroneiki; cv. Mastoides; south Peloponnese; Greece; fatty acids; sterols; botanical origin

---

1. Introduction

Nowadays, the increased awareness regarding the beneficial impact and nutritional properties of extra virgin olive oil (EVOO) is a key factor which has led to a higher demand on international olive oil consumption [1–3]. On the other hand, the increased globalized world and higher cost of olive oil production compared to other vegetable oil sources has led to adulteration with cheaper oils of lower grade. Consequently, a controlled traceability system has become a requirement in the olive oil supply in order to protect consumers against any unapproved and fraudulent practices. Thus, olive oil authenticity and traceability are crucial in order to overcome frauds in the international olive oil trade [4,5]. For this reason, the European Union has adopted a series of regulations in order to certify, protect, and guarantee the quality of the monovarietal olive oils [6–10]. The quality of these monovarietal olive oils is associated with specific characteristics directly related to the olive cultivar [11,12]. Therefore, the authenticity efforts are concentrated on the identification of their botanical origin as well as their adulteration with lower quality or less costly cultivars of lower commercial value.

The production of monovarietal olive oils has increased at a great extent lately since the quality of an olive oil depends on the olive variety from which it originates. Nowadays, several efforts have focused on the investigation of one or several compounds present in olive oils to differentiate olive varieties. Compositional markers include major and minor components providing useful information on olive cultivars to differentiate their botanical origin [11].

Despite the fact that in Greece the number of autochthonous monocultivars is greater than 40, with the most common olive cultivar (for olive oil production) being cv. Koroneiki, the majority of the other autochthonous cultivars remain poorly investigated. Olive cultivation is greatly spread in central Greece, with almost 40% of olive production being centered in the Peloponnese region [13,14]. In southern Peloponnese, among the predominant monovarietal olive oils cultivated are cv. Koroneiki and cv. Mastoides [15]. Koroneiki is the most well-known and systematically cultivated variety, the name of which derives from Koroni, a small village located southeast of Messinia in Peloponnese. On the other hand, Mastoides (referred to locally as Athinolia) is less exploited and cultivated in specific areas of Peloponnese mainly in south Lakonia, Argolida as well as in western Crete. According to our knowledge, there are only two publications for cv. Mastoides, performed in the island of Crete, by Stefanoudaki et al. focusing on the potential of triglyceride and fatty acid composition data as indicators of geographical and botanical origin [16,17].

The present work focuses on the evaluation and characterization of the performances of the two dominant and autochthonous monovarietal olive oils from cv. Koroneiki and cv. Mastoides, cultivated in the south of Peloponnese based on their qualitative and chemical characteristics. Emphasis was given on the influence of cultivar on their fatty acid and sterolic profile in order to be used as compositional/traceability markers in terms of their botanical origin.

2. Materials and Methods

2.1. Geographical Distribution, Sampling, and Sample Maintenance

A total of one hundred and twelve (N = 112) olive oil samples were collected during the harvesting period 2014–2015 from two neighborhood regions in the southern region of Peloponnese in Greece. In particular, sixty nine (69) olive oil samples of Koroneiki cultivar originated from the region of Messinia and forty three (43) olive oil samples of Mastoides cultivar from the southeast part of Lakonia. Both regions are characterized by similar climatic conditions. Olive fruits were picked at the optimal stage of maturity. Samples were transferred to local oil mills in solid, vented, food-grade harvest bins or in suitable harvesting bags. Olive fruits were processed within 24 h, and the same post-harvest conditions were maintained at all cases. In detail, the leaves were removed from the olive fruits, washed and then sent to the crusher. Malaxation was carried out at low temperatures (27–28 °C) for 30–40 min. The obtained olive paste was decanted (horizontal centrifuge) and the resulting olive oil was vertically centrifuged. Olive oil samples were stored directly in 1 L air-tight dark-green glass bottles at 4 °C until further analysis. Quality parameters were analyzed in triplicate, while all the other examined chemical parameters were determined in duplicate.

2.2. Chemicals and Standards

All solvents used for the determination of spectroscopic indices (K_{232}, K_{268}), free fatty acid and peroxide value were purchased from Sigma (St. Louis, MO, USA). The internal standard, 5a-cholestan-3β-ol and the fatty acid methyl esters (FAME) standard mixture were purchased from Sigma (St. Louis, MO, USA). Silica gel plate for thin-layer chromatography was purchased from Fluka (Buchs, Switzerland) and the silylation reagents, pyridine, hexamethyldisilizane, and tri-methylchlorosilane were purchased from Supelco (Bellefonte, PA, USA). Acetone, methanol, n-heptane, chloroform and diethyl-ether were purchased from Sigma (St. Louis, MO, USA).

2.3. Determination of the Physicochemical Quality Parameters

Free fatty acid, peroxide value and spectroscopic indices (K_{232} and K_{268}) were carried out, following the analytical methods described in the Regulation EEC/2568/91 of the European Commission and later amendments [18]. Free fatty acid was expressed as the percentage of oleic acid and peroxide value was given as milliequivalents of active oxygen per kilogram of oil (meq O_2 kg^{-1}). K_{232} and K_{268} extinction coefficients were calculated from absorption at 232 and 268 nm respectively. These absorptions are expressed as specific extinctions E (the extinction of 1% *w/v* solution of the oil in isooctane, in a 10 mm cell) conventionally indicated by K "extinction coefficient".

2.4. Determination of Sterols and Triterpene Dialcohols

The individual sterols, total sterols, and triterpene dialcohols were determined according to the method adopted by EEC/2568/91 regulation, Annexes V with later amendments [18]. The oil sample, with added 5a-cholestan-3β-ol, as an internal standard, was saponified with potassium hydroxide in ethanolic solution and the unsaponifiable matter was extracted with diethyl ether. The sterol and triterpene dialcohol fractions were separated from the unsaponifiable matter by thin-layer chromatography on a basic silica gel plate. The fractions recovered from the silica gel were transformed into trimethylsilyl ethers (TMSE) by the addition of pyridine-hexamethyldisilizane-tri-methylchlorosilane (9:3:1, *v/v/v*). Sterols (%) and triterpene dialcohol contents were determined with a Shimadzu (GC-2010) gas chromatograph equipped with a flame ionization detector (FID), a DB-5 (30 m × 0.32 mm × 0.25 μm) capillary column and an autosampler injector. The operating conditions were as follows: injection temperature 280 °C, column temperature 265 °C, detector temperature 310 °C, splitting ratio (1:50), flow rate 1.4 mL/min, and injection volume of 1 μL of TMSE solution. Individual sterols were identified based on their relative retention times with respect to the internal standard, 5a-cholestan-3β-ol, according to the standardized reference method [18]. The sterols and triterpene dialcohols eluted in the following order: cholesterol, 24-methylen-cholesterol, campesterol, campestanol, stigmasterol, Δ7-campesterol, Δ5,23-stigmastadienol, clerosterol, β-sitosterol, sitostanol, Δ5-avenasterol, Δ5,24-stigmastadienol, Δ7-stigmastenol, Δ7-avenasterol, erythrodiol, and uvaol (calculated as total erythrodiol). Sterol and triterpene diol concentrations were calculated as mg/kg of oil with respect to the internal standard. Results were expressed as proportions (%) of total sterols. The sum of Δ5,23-stigmastadienol, clerosterol, β-sitosterol, sitostanol, Δ5-avenasterol, and Δ5,24-stigmastadienol represents apparent b-sitosterol. Mean values of duplicate experiments in each sample were used for further statistical analysis.

2.5. Determination of Fatty Acid Composition

The fatty acid composition was determined according to the official method of the Regulation EEC/2568/91, Annex IV with amendments [18]. The fatty acid methyl esters (FAME) were obtained by cold alkaline transesterification with methanolic potassium hydroxide solution and extracted with *n*-heptane. FAME were analyzed on a model GC-2010 Shimadzu chromatograph, equipped with a BPX-70, (60 m × 0.25 mm × 0.25 μm), capillary column and a flame ionization detector (FID). The carrier gas was helium, with a flow of 1.5 mL/min. The temperatures of the injector and detector were set at 250 and 260 °C respectively and the oven temperature was increased gradually from 165 to 225 °C in 35 min. The injection volume was 1 μL. Quantification was achieved using a FAME standard mixture. The results were expressed as a percentage of individual fatty acids.

2.6. Statistical and Chemometric Analysis

Results were expressed as mean values ± standard deviation (SD). Data was processed with MINITAB 18 software. Thus, it was possible to obtain the minimum and the maximum value of the sample, mean, and standard deviation (SD). The minimum and the maximum value of the sample are the values of the largest and smallest elements of a sample. In statistics, the difference between the largest and smallest values (range) provides an indication of statistical dispersion. Differences between means were tested for statistical significance using analysis of variance (ANOVA). Statistical significance level was set at $p < 0.05$. In addition, principal component analysis (PCA) was applied to study the relations between the two (2) mono-cultivars (cv. Koroneiki vs cv. Mastoides) on the examined chemical properties.

3. Results and Discussion

3.1. Physico-Chemical Parameter of the Two Major Olive Cultivars of Southern Peloponnese

It is well established that the quality parameters of olive oil are mainly altered by factors causing injuries to the olive fruits such as olivefly attacks, improper methods during olive harvesting as well as poor post-extraction conditions (e.g., inappropriate storage and packaging) [19]. Free fatty acid, peroxide value and spectrophotometric absorption were examined in the studied olive oils. Mean values for each analytical parameter, as well as minimum and maximum values of the measured parameters are reported in Table 1. It is clear that all analysed samples obtained from the two examined cultivars in the southern region of Peloponnese (cv. Koroneiki and cv. Mastoides), are classified in the highest quality category as "extra virgin olive oil (EVOO)" as they fullfil the demands of the current EU Regulation 2568/91 [18]. In particular, olive oils of both cultivars exhibited low mean values on their qualitative parameters. The mean free fatty acid was 0.34% for olive oils of cv Koroneiki and 0.39% for olive oils of cv. Mastoides. Respectively, the mean peroxide value for cv. Koroneiki was 7.24 meqO$_2$ kg^{-1} and 6.96 meqO$_2$ kg^{-1} for olive oils of cv. Mastoides. Likewise, K$_{232}$ and K$_{268}$ mean values were quite below the limit set by the EU Regulation 2568/91. The results depict the overall high quality of south Peloponesse olive oil production, one of the most important olive-growing regions in Greece.

Table 1. Qualitative parameters from the two major olive cultivars of southern Peloponnese.

	cv. Koroneiki (N = 69)		cv. Mastoides (N = 43)		EEC Limit for EVOO Category
Parameter	Mean ± SD	Min–Max	Mean ± SD	Min–Max	
Free acidity (%)	0.34 ± 0.13	0.17–0.76	0.39 ± 0.13	0.15–0.77	≤0.80
Peroxide value (meqO$_2$/kg)	7.24 ± 1.88	3.64–11.96	6.96 ± 2.31	2.88–14.70	≤20
K$_{232}$	1.55 ± 0.14	1.33–2.14	1.63 ± 0.11	1.33–2.02	≤2.50
K$_{268}$	0.13 ± 0.01	0.08–0.21	0.12 ± 0.02	0.08–0.17	≤0.22

Results are expressed as means ± standard deviation (SD). N = 112.

3.2. Evaluation and Discrimination of the Two Examined Cultivars of Southern Peloponnese According to Their Fatty Acid Composition

Fatty acid composition is a crucial parameter for the quality and characterization of an olive oil [20]. Because of the fact that the fatty acid content is a fundamental parameter for the determination of the nutritional properties of olive oil, the description of a specific cultivar on the basis of their fatty acid composition is of utmost importance. As a result, many researchers have used fatty acid composition in order to group olive oils according to the origin of the cultivar [21–23].

In the present study, the GC-FID analysis of the 112 olive oil samples from Koroneiki and Mastoides cultivars showed their complete fatty acid composition. As shown in Table 2, all values of the thirteen fatty acids identified, were in conformity to the normal range expected for olive oil category for both cultivars. Generally, olive oils of Koroneiki cultivar had a mean value of 76.70% for the mono-unsaturated oleic acid (C18:1) compared to olive oils of Mastoides cultivar which had a mean value of 75.93% ($p < 0.05$). Moreover, olive oils of Koroneiki presented a higher concentration with respect to the poly-unsaturated linolenic acid (C18:3) with a mean value of 0.68% compared to cv. Mastoides (0.55%). On the other hand, olive oils of cv. Mastoides were characterized by a clearly higher concentration in heptadecanoic acid (C17:0) with a mean value at 0.14% and in heptadecenoic acid (C17:1) with a mean value at 0.25% compared to the olive oils of cv. Koroneiki which had almost a three-fold lower concentration, with mean values 0.05% and 0.08%, respectively. No differences were observed for the following fatty acids: myristic (C14:0), palmitic (C16:0), palmitoleic (C16:1), and linoleic acids (C18:2) as shown in Table 2.

Table 2. Fatty acid profile of cv. Koroneiki and cv. Mastoides cultivated in southern Peloponnese.

Parameter	cv. Koroneiki (N = 69)		cv. Mastoidis (N = 58)		Calculated *p*-Value	EEC Limit for EVOO Category
	Mean ± SD	Min–Max	Mean ± SD	Min–Max		
Myristic C14:0 (%)	0.01 ± 0.00	0.00–0.02	0.01 ± 0.00	0.00–0.02	n.s	≤0.03
Palmitic C16:0 (%)	12.02 ± 0.74	9.54–13.56	12.29 ± 0.77	9.96–13.28	n.s	7.50–20.00
Palmitoleic C16:1 (%)	0.92 ± 0.13	0.64–1.43	0.92 ± 0.10	0.64–1.08	n.s	0.30–3.50
Heptadecanoic C17:0 (%)	0.05 ± 0.02	0.03–0.15	0.14 ± 0.02	0.08–0.17	0.00	≤0.40
Heptadecenoic C17:1 (%)	0.08 ± 0.04	0.06–0.24	0.25 ± 0.03	0.16–0.29	0.00	≤0.60
Stearic C18:0 (%)	2.53 ± 0.19	1.98–3.12	2.64 ± 0.16	2.35–2.99	0.001	0.50–5.00
Oleic C18:1 (%)	76.70 ± 1.96	70.67–81.40	75.93 ± 1.27	73.15–79.44	0.024	55.00–83.00
Linoleic C18:2 (%)	6.09 ± 1.60	4.20–12.01	6.44 ± 0.69	5.11–8.13	n.s	2.50–21.00
Linolenic C18:3 (%)	0.68 ± 0.07	0.51–0.86	0.55 ± 0.04	0.49–0.66	0.00	≤1.00
Arachidic C20:0 (%)	0.44 ± 0.03	0.33–0.50	0.39 ± 0.02	0.35–0.44	0.00	≤0.60
Eicosenoic C20:1 (%)	0.31 ± 0.02	0.27–0.35	0.27 ± 0.02	0.23–0.32	0.00	≤0.50
Behenic C22:0 (%)	0.14 ± 0.01	0.09–0.17	0.10 ± 0.01	0.07–0.12	0.00	≤0.20
Lignoceric C24:0 (%)	0.05 ± 0.00	0.03–0.08	0.04 ± 0.007	0.03–0.06	0.00	≤0.20

Results are expressed as means ± standard deviation (SD). n.s = not-significant. The statistical significance level was set at $p < 0.05$.

As mentioned in the introduction, there is one relevant publication by Stefanoudaki et al. [17] where the authors examined the same cultivars in the island of Crete. They reported that olive oils of Koroneiki cultivar were characterized by lower concentrations of oleic (C18:1) and heptadecanoic acids (C17:0) and higher concentrations of linoleic (C18:2) and palmitic acids (C16:0). Those differences can be explained by the fact that apart from the olive cultivar other secondary factors, mainly environmental, (e.g., different climatic conditions such as temperature, rainfall, humidity at each growing site), have a significant effect on the composition of the fatty acid profile [24,25].

As shown in Table 2, the fatty acid composition data of the 112 olive oil samples were subjected to analysis of variance. It was revealed that, apart from C14:0, C16:0, C16:1, and C18:2, substantial differences were observed between Koroneiki and Mastoides cultivars in all the rest analyzed fatty acids ($p < 0.05$). Additionally, principal component analysis (PCA) on fatty acid composition data was performed to confirm and enhance the classification according to the cultivar. PCA can be used to decrease the initial variables into a limited number of new variables (principal components) describing most of the variation in the originals. The main purpose of the key factor analysis, taken together, is to define related variables. The first two principal components are significant and explain approximately the 84% of the variation in the data. Thus, based on PCA in Figure 1 we showed the score plot of PCA for cv. Koroneiki and cv. Mastoides according to their fatty acid composition. In this case we found that most of the points for K are pointed to the left of PC1, meaning that K has large negative loadings on component 2. On the other hand, points for M are presented on the right of PC1, meaning that M has large positive loadings on component 1. The K and M regions are therefore independent of each other and the chemical properties studied are also independent and the regions are affected by them. Hence, the application of the PCA algorithm to the fatty acid data revealed a discrete separation between the two cultivars, by creating two distinctive clusters. The results are in agreement with studies by Stefanoudaki et al. where they concluded that fatty acid compositional data of Koroneiki and Mastoides cultivar showed significant potential for olive oil classification [17].

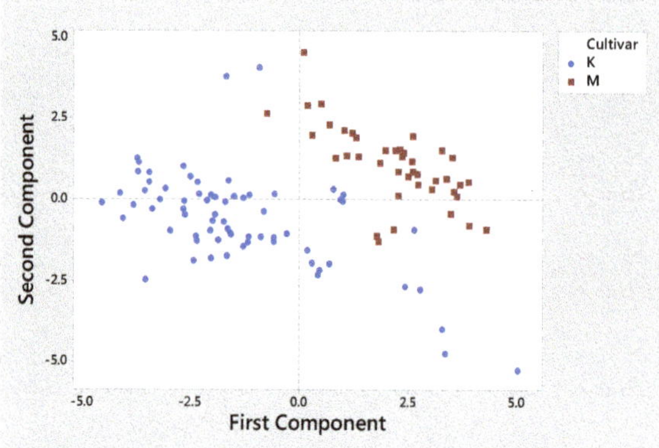

Figure 1. Score plot of principal component analysis (PCA) for cv. Koroneiki and cv. Mastoides obtained from olive trees in southern Peloponnese according to their fatty acid composition. K corresponds to Koroneiki olive oils (blue dots) and M to Mastoides olive oils (red dots).

There are other studies focusing on olive oil compounds with the capability to differentiate among cultivars, highlighting that fatty acid composition data can be used as a traceability marker of the botanical origin [21,26–28]. For example, D' Imperio et al. by analyzing Sicilian extra virgin olive oils from 22 cultivars found out that oleic, linoleic and palmitic fatty acids were crucial in the characterization of the olive oil cultivars [21]. Likewise, Krichene et al. determined the content of fatty acids and phenolic compounds, as well as other olive oil minor components in Tunisian olive cultivars; observing clear differences between them [28].

3.3. Evaluation and Discrimination of the Two Examined Cultivars From the Southern Region of Peloponnese According to Their Sterolic Profile

Olive oil is characterized by several minor components with an important nutritional impact on human health [29,30]. Phytosterols and triterpenic dialcohols are included among them and constitute the major proportion of the unsaponifiable fraction of olive oil (around 20%). Many researchers have revealed that the application of different chemometric treatments on the sterols present in olive oils or a combination of specific individual sterols with other chemical parameters can discriminate among olive cultivars [31–35]. For example Lukic et al. demonstrated that sterols and triterpene diols can be used as reliable indicators of variety and ripening degree among virgin olive oils from Croatia [31]. Another research group has shown that the combination of total sterol content, campesterol, stearic acid, and oxidative stability enabled the classification of olive oils according to their variety [36].

Although several studies have been conducted for Greek mono-cultivars in other regions of Greece [37–40], in the present study, the sterolic composition and content from the two monovarietal olive oils of southern Peloponnese were evaluated and compared. Table 3 lists the mean values expressed as percentage of the individual sterols and total sterols concentration of the two monocultivars. The individual sterols and total sterols content for the examined olive oil sample of Mastoides cultivar were within the established EU regulatory limits [18]. In general, Mastoides oils exhibited higher mean value for β-sitosterol (84.12%) and lower mean value for Δ-5-avenasterol (9.85%) and total erythodiol content (1.40%) compared to the relative values for Koroneiki olive oils (Table 3). In addition, higher concentration in the mean total sterols was observed in Mastoides olive oils (1219.6 mg/kg) compared to the olive oils of Koroneiki cultivar, where the mean value was 1033.3 mg/kg, very close to the regulatory set limit of 1000 mg/kg according to the EU regulation 2568/91 [41].

Table 3. Sterol profile of cv. Koroneiki and cv. Mastoides cultivated in southern Peloponnese.

	cv. Koroneiki (N = 69)	cv. Mastoidis (N = 43)	Calculating *p*-Value	EEC Limit for EVOO Category
Sterols and Triterpene Diols	**Mean ± SD**	**Mean ± SD**		
Cholesterol (%)	0.11 ± 0.03	0.12 ± 0.03	0.017	≤0.5
24-methylene-cholesterol %	0.32 ± 0.09	0.19 ± 0.05	0.00	
Campesterol %	3.71 ± 0.38	3.14 ± 0.16	0.00	≤4.0
Campestanol %	0.05 ± 0.03	0.04 ± 0.02	n.s	<campesterol
Stigmasterol %	0.74 ± 0.19	0.64 ± 0.18	0.01	
Chlerosterol %	0.85 ± 0.07	0.94 ± 0.07	0.00	
β-Sitosterol %	80.73 ± 3.73	84.12 ± 2.69	0.00	
Sitostanol %	0.37 ± 0.30	0.31 ± 0.08	n.s	
Δ-5-avenasterol %	12.28 ± 3.96	9.85 ± 2.66	0.001	
Δ-5, 24-stigm/dienol %	0.29 ± 0.10	0.22 ± 0.06	0.00	
Δ-7-stigmastenol %	0.19 ± 0.09	0.18 ± 0.09	n.s	≤0.5
Δ-7-avenasterol %	0.28 ± 0.11	0.22 ± 0.06	0.001	
Apparent b-Sitosterol %	94.63 ± 1.07	95.45 ± 0.29	0.00	≥93.0
Total Erythrodiol %	2.85 ± 1.25	1.40 ± 0.52	0.00	≤4.5
Total sterols (mg/kg)	1033.3 ± 150.1	1219.6 ± 109.2	0.00	≥1000

Results are expressed as means ± standard deviation (SD). n.s = not-significant. The statistical significance level was set at $p < 0.05$.

As shown in Table 3, by comparing the two cultivars, the calculated *p*-value according to their sterolic profile, was in most cases close to 0.00 ($p \approx 0.00$), indicating a strong botanical effect. No previous reported data is available to compare and to the best of our knowledge, it is the first time to examine the sterolic profile of cv. Mastoides. The PCA score plot of Koroneiki versus Mastoides olive oils according to their sterolic profile is presented in Figure 2. The first two principal components explain approximately the 81% of the variation in the data. It is observed that most of the K points are shown to the right of PC1, hence K has large positive loadings on component 1. On the other hand, most of the M points have large negative loadings on component 2. Thus, K and M regions are independent of each other according to their sterolic profile, permitting a clear classification of the examined monocultivars in two separated clusters. Relevant studies in Greek olive cultivars have been carried out classifying Greek olive oils according to cultivar and geographical origin, based on the composition of their volatile compounds [42], phenolic compounds and fatty acids composition [38].

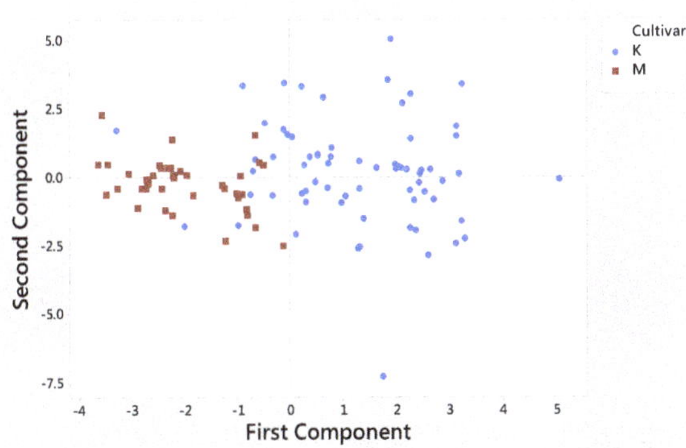

Figure 2. Score plot of PCA for cv. Koroneiki and cv. Mastoides according to their sterolic profile. K corresponds to Koroneiki olive oils (blue dots) and M to Mastoides olive oils (red dots).

According to many authors, chemometric tools can also be used to select the best variables to obtain satisfactory results [32–34,36,43,44]. As a result, a combined principal component analysis was performed using both fatty acid compositional data and individual/total sterols as variables. To simplify the method used to limit a large set of variables to a small set but holding most of the detail in the large set, PCA was applied in this case too. The first two principal components illustrate data variation of 81%. The score plot of PCA for cv. Koroneiki and cv. Mastoides according to the combination of fatty acid compositional data and sterolic profile is shown in Figure 3. In this scenario, we found that the majority of the points for K stand on the left side of PC1, and hence K has large negative loads on component 2. On the other side, most of the points for M stand on the right of PC1, thus implying M has large positive loads at component 1. Both K and M regions are therefore independent of each other according to the combination of fatty acid and sterolic profile.

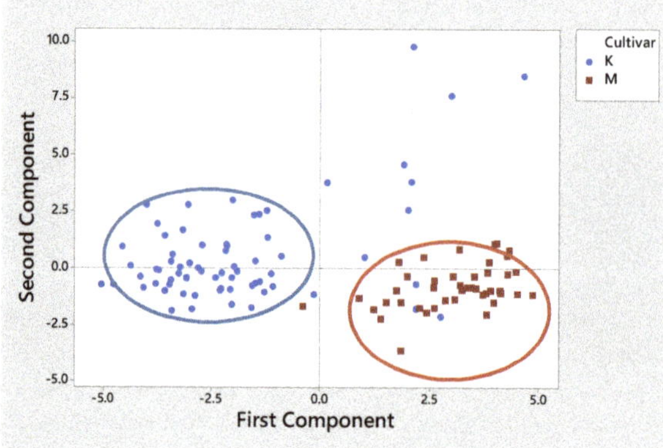

Figure 3. Score plot of PCA for cv. Koroneiki and cv. Mastoides olive oil samples of southern Peloponnese according to the combination of fatty acid compositional data and sterolic profile. K corresponds to Koroneiki olive oils (blue dots) and M to Mastoides olive oils (red dots).

Thus, it is evident to conclude that fatty acid and sterolic profile data can permit the discrimination of the examined extra virgin olive oils in south Peloponnese region in terms of olive cultivar and can be used as useful authenticity-traceability indicators.

4. Conclusions

In the present study we demonstrated that fatty acid compositional data and sterols have a high differentiation potential as authenticity tools. Meanwhile, analyses on other more or less exploited Greek monocultivars need to be performed in order to reveal and evaluate their quality and chemical characteristics so as to establish a national authenticity databank. Finally, the possibility of investigating other components present in olive oils and taking into account new authenticity methodologies would be useful for the comparison of different Greek monocultivars in the region.

Author Contributions: V.S. designed and performed the experiments; P.T. performed statistical analysis; T.V. and V.S. wrote, edited, and reviewed the paper. All authors have read and agreed to the published version of the manuscript.

Funding: This research project was funded under the Project "Research and Technology Development Innovation Projects"-AgroETAK, MIS 453350, in the framework of the Operational Program "Human Resources Development." It is co-funded by the European Social Fund through the National Strategic Reference Framework (Research Funding Program 2007–2013) coordinated by the Hellenic Agricultural Organization–DEMETER.

Acknowledgments: V.S. would like to thank A.S. Vekiari and the Dept. of Olive and Horticultural crops, Kalamata, Greece, for the support provided. All the experiments were conducted in the accredited laboratory of the Dept. of Olive and Horticultural crops, Kalamata, Greece, according to ISO 17025:2008.

Conflicts of Interest: The authors declare no conflict of interest.

References

1. International Olive Council Database. Available online: http://www.internationaloliveoil.org/estaticos/view/131-world-olive-oil-figures (accessed on 17 November 2019).
2. Prospects for the Olive Oil Sector in Spain, Italy and Greece 2012–2020. Available online: https://ec.europa.eu/agriculture/markets-and-prices/market-briefs_en (accessed on 10 November 2019).
3. Zampounis, V. Olive oil in the World Market. In *Olive Oil: Chemistry and Technology*, 2nd ed.; Boskou, D., Ed.; AOCS Press: Campaign, IL, USA, 2006; pp. 21–40.
4. Montet, D.; Ray, R. (Eds.) *Food Traceability and Authenticity*; CRC Press: Boca Raton, FL, USA, 2018. [CrossRef]
5. Conte, L.; Bendini, A.; Valli, E.; Lucci, P.; Moret, S.; Maquet, A.; Lacoste, F.; Brereton, P.; Garcia-Gonzalez, D.L.; Moreda, W.; et al. Olive oil quality and authenticity: A review of current EU legislation, standards, relevant methods of analyses, their drawbacks and recommendations for the future. *Trends Food Sci. Technol.* **2019**, in press. [CrossRef]
6. Council Regulation (EC). No. 2081/92 of 14 July 1992 on the protection of geographical indications and designations of origin for agricultural products and foodstuffs. *Off. J. Eur. Union* **1992**, *L208*, 1–8.
7. Council Regulation (EC). No. 2082/92 of 14 July 1992 on certificates of specific character for agricultural products and foodstuffs. *Off. J. Eur. Union* **1992**, *L208*, 9–14.
8. Council Regulation (EC). No. 510/2006 of 20 March 2006 on the protection of geographical indications and designations of origin for agricultural products and foodstuffs. *Off. J. Eur. Union* **2006**, *L93*, 12–25.
9. Council Regulation (EC). No. 1898/2006 of 14 December 2006 laying down detailed rules of implementation of Council Regulation (EC) no. 510/2006 on the protection of geographical indications and designations of origin for agricultural products and foodstuffs. *Off. J. Eur. Union* **2006**, *L369*, 1–23.
10. Council Regulation (EC). No. 182/2009 of 6 March 2009 amending Regulation (EC) no. 1019/2002 on marketing standards for olive oil. *Off. J. Eur. Union* **2009**, *L63*, 6–8.
11. Montealegre, C.; Alegre, M.; García-Ruiz, C. Traceability Markers to the Botanical Origin in Olive Oils. *J. Agric. Food Chem.* **2009**, *58*, 28–38. [CrossRef]
12. Aparicio, R.; Luna, G. Characterization of monovarietal virgin olive oil. *J. Lipid Sci. Technol.* **2002**, *104*, 614–627. [CrossRef]
13. Olive Oil: Establishing the Greek Brand, National Bank of Greece. Available online: https://www.nbg.gr/greek/the-group/press-o_ce/e-spot/reports/Documents/Olive%20Oil_2015.pdf (accessed on 23 October 2019).
14. Kostalenos, G. Greek olive varieties. In *Elaiokomia Egkiklopedia-The Olive Oil*, 1st ed.; Zambounis, V., Ed.; Axion Ekdotiki: Athens, Greece, 2017; pp. 39–53. (In Greek)
15. Kostalenos, G. *Olive Fruit Data, History, Description and Geographical Distribution of Olive Varieties in Greece*, 1st ed.; Kostalenos: Poros Trizinias, Greece, 2011; pp. 1–436. (In Greek)
16. Stefanoudaki, E.; Kotsifaki, F.; Koutsaftakis, A. The potential of HPLC triglyceride profiles for the classification of Cretan olive oils. *Food Chem.* **1997**, *60*, 425–432. [CrossRef]
17. Stefanoudaki, E.; Kotsifaki, F.; Koutsaftakis, A. Classification of virgin olive oils of the two major Cretan cultivars based on their fatty acid composition. *J. Am. Oil Chem. Soc.* **1999**, *76*, 623–626. [CrossRef]
18. Commission Regulation (EEC). No. 2568/91 of 14 July 1991 on the characteristics of olive oil and olive-residue oil and on the relevant methods of analysis. *Off. J. Eur. Union* **1991**, *L208*, 1–8.
19. Kiritsakis, A.; Christie, W. Analysis of edible oil. In *Handbook of Olive Oil*; Harwood, J., Aparicio, R., Eds.; Springer: Aspen, CO, USA; Gaithersburg, MD, USA, 2000; pp. 129–158.
20. Boskou, D. Olive oil. In *More on Mediterranean Diets*; Simopoulos, A.P., Visioli, F., Eds.; Karger: Basel, Switzerland, 2007; pp. 180–210.
21. D'Imperio, M.; Dugo, G.; Alfa, M.; Mannina, L.; Segre, A.L. Statistical analysis on Sicilian olive oils. *Food Chem.* **2007**, *102*, 956–965. [CrossRef]
22. Sanchez Casas, J.J.; Osorio Bueno, E.; Montano Garcia, A.M.; Martinez Cano, M. Estudio del contenido en acidos grasos de aceites monovarietales elaborados a partir de aceitunas producidas en la region extremena. *Grasas y Aceites* **2003**, *54*, 371–377.

23. Mannina, L.; Dugo, G.; Salvo, F.; Cicero, L.; Ansanelli, G.; Calcagni, C.; Segre, A. Study of the cultivar-composition relationship in Sicilian olive oils by GC, NMR, and statistical methods. *J. Agric. Food Chem.* **2003**, *51*, 120–127. [CrossRef] [PubMed]
24. Mailer, R.J.; Ayton, J.; Graham, K. The Influence of growing region, cultivar and harvest timing on the diversity of Australian olive oil. *J. Am. Oil Chem. Soc.* **2010**, *87*, 877–884. [CrossRef]
25. Longobardi, F.; Ventrella, A.; Casielloa, G.; Sacco, D.; Tasioula-Margari, A.K.; Kiritsakis, A.K.; Kontominas, M.G. Characterisation of the geographical origin of Western Greek virgin olive oils based on instrumental and multivariate statistical analysis. *Food Chem.* **2012**, *133*, 1169–1175. [CrossRef]
26. Di Bella, G.; Maisano, R.; La Pera, L.; Lo Turco, V.; Salvo, F.; Dugo, G. Statistical characterization of Sicilian olive oils from the Peloritana and Maghrebian zones according to the fatty acid profile. *J. Agric. Food Chem.* **2007**, *55*, 6568–6574. [CrossRef]
27. Ollivier, D.; Artaud, J.; Pinatel, C.; Durbec, J.P.; Guerere, M. Triacylglycerol and fatty acid compositions of French virgin olive oils. Characterization by chemometrics. *J. Agric. Food Chem.* **2003**, *51*, 5723–5731. [CrossRef]
28. Krichene, D.; Taamalli, W.; Daoud, D.; Salvador, M.; Fregapane, G.; Zarrouk, M. Phenolic compounds, tocopherols and other minor components in virgin olive oils of some Tunisian varieties. *J. Food Biochem.* **2007**, *31*, 179–194. [CrossRef]
29. Boskou, D. *Olive Oil: Minor. Constituents and Health*; CRC Press: Boca Raton, FL, USA, 2010.
30. Leontopoulos, S.; Skenderidis, P.; Kalorizou, H.; Petrotos, K. Bioactivity Potential of Polyphenolic Compounds in Human Health and their Effectiveness Against Various Food Borne and Plant Pathogens. A Review. *Int. J. Food Biol. Eng.* **2017**, *7*, 1–19.
31. Lukic, M.; Lukic, I.; Krapac, M.; Sladonja, B.; Pilizota, V. Sterols and triterpene diols in olive oil as indicators of variety and degree of ripening. *Food Chem.* **2013**, *136*, 251–258. [CrossRef] [PubMed]
32. Bucci, R.; Magri, A.D.; Magri, A.L.; Marini, D.; Marini, F. Chemical authentication of extra virgin olive oil varieties by supervised chemometric procedures. *J. Agric. Food Chem.* **2002**, *50*, 413–418. [CrossRef] [PubMed]
33. Matos, L.C.; Cunha, S.C.; Amaral, J.S.; Pereira, J.A.; Andrade, P.B.; Seabra, R.M.; Oliveira, B. Chemometric characterization of three varietal olive oils (cvs. Cobrancosa, Madural and Verdeal Transmontana) extracted from olives with different maturation indices. *Food Chem.* **2007**, *102*, 406–414. [CrossRef]
34. Brescia, M.A.; Alviti, G.; Liuzzi, V.; Sacco, A. Chemometrics classification of olive cultivars based on compositional data of oils. *J. Am. Oil Chem. Soc.* **2003**, *80*, 945–950. [CrossRef]
35. Nagy, K.; Bongiorno, D.; Avellone, G.; Agozzino, P.; Ceraulo, L.; Vekey, K. High performance liquid chromatography-mass spectrometry based chemometric characterization of olive oils. *J. Chromatogr. A* **2005**, *1078*, 90–97. [CrossRef]
36. Lorenzo, I.M.; Perez Pavon, J.L.; Fernandez Laespada, M.E.; Garcıa Pinto, C.; Moreno Cordero, B.; Henriques, L.R.; Peres, M.F.; Simoes, M.P.; Lopes, P.S. Application of headspace-mass spectrometry for differentiating sources of olive oils. *Anal. Bioanal. Chem.* **2002**, *374*, 1205–1211. [CrossRef]
37. Vekiari, S.A.; Oreopoulou, V.; Kourkoutas, Y.; Kamoun, N.; Msallem, M.; Psimouli, V.; Arapoglou, D. Characterization and seasonal variation of the quality of virgin olive oil of the Throumbolia and Koroneiki varieties from Southern Greece. *Grasas y Aceites* **2010**, *61*, 221–231. [CrossRef]
38. Agiomirgianaki, A.; Petrakis, P.; Dais, P. Influence of harvest year, cultivar and geographical origin on Greek extra virgin olive oils composition: A study by NMR spectroscopy and biometric analysis. *Food Chem.* **2012**, *135*, 2561–2568. [CrossRef]
39. Karabagias, I.; Michos, C.; Badeka, A.; Kontakos, S.; Stratis, I.; Kontominas, M.G. Classification of Western Greek virgin olive oils according to geographical origin based on chromatographic, spectroscopic, conventional and chemometric analyses. *Food Res. Int.* **2013**, *54*, 1950–1958. [CrossRef]
40. Kandylis, P.; Vekiari, A.S.; Kanellaki, M.; Grati Kamoun, N.; Msallem, M.; Kourkoutas, Y. Comparative study of extra virgin olive oil flavor profile of Koroneiki variety (Olea europaea var. Microcarpa alba) cultivated in Greece and Tunisia during one period of harvesting. *LWT J. Food Sci. Technol.* **2011**, *44*, 1333–1341. [CrossRef]
41. Skiada, V.; Tsarouhas, P.; Varzakas, T. Preliminary study and observation of "Kalamata PDO" extra virgin olive oil, in the Messinia region, southwest of Peloponnese (Greece). *Foods* **2019**, *8*, 610. [CrossRef] [PubMed]

42. Pouliarekou, E.; Badeka, A.; Tasioula-Margaria, M.; Kontakos, S.; Longobardic, F.; Kontominas, M.G. Characterization and classification of Western Greek olive oils according to cultivar and geographical origin based on volatile compounds. *J. Chromatogr. A* **2011**, *1218*, 7534–7542. [CrossRef] [PubMed]
43. Aranda, F.; Gomez-Alonso, S.; Rivera del _Alamo, R.M.; Salvador, M.D.; Fregapane, G. Triglyceride, total and 2-position fatty acid composition of Cornicabra virgin olive oil: Comparison with other Spanish cultivars. *Food Chem.* **2004**, *86*, 485–492. [CrossRef]
44. Lerma-Garcıa, M.J.; Herrero-Martınez, J.M.; Ramis-Ramos, G.; Simo-Alfonso, E.F. Prediction of the genetic variety of Spanish extra virgin olive oils using fatty acid and phenolic compound profiles established by direct infusion mass spectrometry. *Food Chem.* **2008**, *108*, 1142–1148. [CrossRef] [PubMed]

 *foods*

Article

Preliminary Study and Observation of "Kalamata PDO" Extra Virgin Olive Oil, in the Messinia Region, Southwest of Peloponnese (Greece)

Vasiliki Skiada [1,*], Panagiotis Tsarouhas [2] and Theodoros Varzakas [1,*]

[1] Department of Food Science and Technology, University of Peloponnese, Antikalamos, 24100 Kalamata, Greece
[2] Department of Supply Chain Management (Logistics), International Hellenic University, Kanellopoulou 2, 60100 Katerini, Greece; ptsarouh@mail.teiste.gr
* Correspondence: vpskiada@yahoo.gr (V.S.); theovarzakas@yahoo.gr (T.V.); Tel.: +30-27210-45279 (T.V.)

Received: 4 November 2019; Accepted: 19 November 2019; Published: 23 November 2019

Abstract: While there has been considerable research related to Koroneiki cultivar in different areas in Greece, no systematic work has been carried out on olive oil analysis from one of the most important olive-growing regions in Greece, located southwest of Peloponnese, Messinia. This work is the first systematic attempt to study the profile of Messinian olive oils and evaluate to what extent they comply with the recent EU regulations in order to be classified as "Kalamata Protected Designation of Origin (PDO)"-certified products. Quality indices were measured and detailed analyses of sterols, triterpenic dialcohols, fatty acid composition and wax content were conducted in a total of 71 samples. Messinian olive oils revealed a high-quality profile but, at the same time, results demonstrated major fluctuations from the established EU regulatory limits on their chemical parameters. Results showed low concentrations of total sterols, with 66.7% of the examined samples below the regulated set limits for Kalamata PDO status; high concentrations of campesterol, with a total of 21.7%, exceeding the legal maximum of 4.0%; and a slight tendency of high total erythrodiol content. Fatty acid composition and wax content were within the normal range expected for the extra virgin olive oil (EVOO) category. However, the narrower established PDO limits in specific fatty acids showed some fluctuations in a few cases.

Keywords: EVOO; Kalamata PDO; Koroneiki cultivar; Greece; Messinia region; EU regulations; quality and chemical parameters; sterols

---

1. Introduction

Olive oil is a key element of the Mediterranean diet as well as an exceptional lipid source. Prestigious scientific studies have acknowledged olive oil as a healthy food with multiple utilities in, and benefits for, the human body [1,2]. Nowadays, it is well established that the health-promoting effects of extra virgin olive oil are attributed not only due to its high oleic acid content but also due to its unique bioactive polar phenolic compounds [3–5]. As a result, the biological properties, health-promoting effects and nutritive characteristics of extra virgin olive oil have led to a continuous growth in its consumption [6].

Greece is ranked third among olive oil-producing countries, after Spain and Italy, with approximately 16% of the annual production worldwide. Almost 60% of Greece's arable land is taken up by olive trees. It is the world's top producer of black olives and has more olive cultivars than any other country worldwide. The annual olive oil production is approximately 300,000–400,000 tons, depending on the harvest year, and 80% of the olive oil produced belongs to the category of extra virgin olive oil (EVOO) [7–9]. Hence, olive cultivation in Greece represents not only a crucial resource for rural economies but also an important part of the social, cultural and environmental heritage, as more than 450,000 families work in the fields of olive cultivation [10].

Geographically speaking, almost 70% of olive oil production in Greece is centered in two regions—Peloponnese (39%) and Crete (30%)—with the prefecture of Messinia being the dominant olive-growing area of Peloponnese [10]. Koroneiki cultivar (Olea europeae var. Microcarpa alba) is the indigenous variety in Messinia—the name of which derives from Koroni, a small seaside village southeast of Messinia [11].

Although there are many research publications related to Koroneiki cultivar in different areas in Greece [12–17], no systematic work has been carried out on olive oil analysis from the Messinia region. In August 2015, the European Commission approved the extension of the "Kalamata Protected Designation of Origin (PDO) olive oil" from the former province of Kalamata to the rest Regional Unit of Messinia, considerably enlarging the area covered by the PDO [18–22]. On this basis, the new "Kalamata PDO olive oil" introduces more stringent criteria/specifications than those laid down in the European Commission Regulation 2568/1991 for extra virgin oil in order to ensure that the name "Kalamata PDO olive oil" is used only for the area's olive oil [21,22]. This recent approval, throughout the boundaries of Messinia, could be a very competitive advantage with an important added value, giving a higher market price and a robust commercial presence to "Kalamata PDO Olive oil," as a PDO trademark is considered an additional guarantee of quality, authenticity, tradition and safety [23–25]. However, it is questionable whether Messinian olive oils meet the requirements of the "Kalamata PDO olive oil" profile.

The aim of this study was to investigate, evaluate and report the qualitative and chemical parameters of extra virgin olive oils obtained from the Messinia region. This data will be a useful and important tool in profiling their typical characteristics and evaluating the extent to which they comply with the amended regulation in order to be classified as PDO-certified products. Finally, this study is a motivation for a deeper investigation of the Messinian olive oil, from the southwest region of Peloponnese, which is one of the most important olive-growing regions in Greece and, at the same time, very little investigated.

2. Materials and Methods

2.1. Geographical Distribution and Selection of Olive Oil Samples

A total of seventy-one (71) olive oil samples were obtained in one successive harvesting year (2014–2015), cultivated in the geographical Messinia region, southwest of Peloponnese, in Greece (see Figure 1). All samples were produced from olive trees representing the typical Koroneiki cultivar of Messinia. Sampling was made from different points in the prefecture of Messinia so as to have the utmost homogeneity. As mentioned earlier, the European Commission recently approved the extension of "Kalamata PDO olive oil" throughout the whole Messinia region, enlarging the area covered by the PDO. On this basis, Messinian extra virgin olive oils may be classified as PDO, if they meet the corresponding parameters [21]. The whole region is characterized by the same climatic conditions as described in the relevant EC Commission Regulation for Kalamata PDO olive oil [19–21].

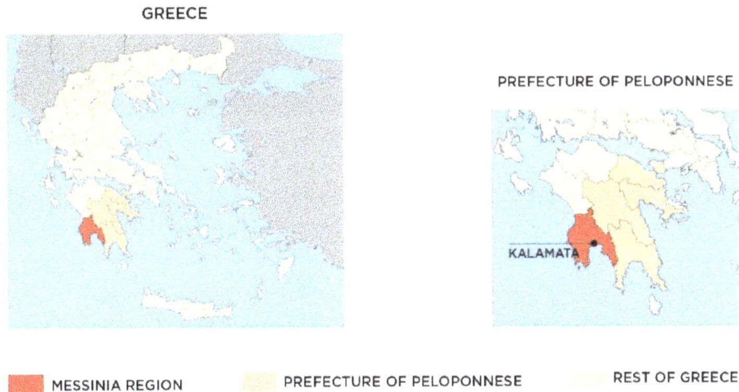

Figure 1. Map of Greece, focusing on the Messinia region (in red), southwest of the Prefecture of Peloponnese (in orange). Adapted from Wikipedia [22].

2.2. Sampling and Sample Maintenance

Sampling was carried out during the 2014–2015 olive fruit-harvesting period. Provision was made to harvest olive fruits at the optimal stage of maturity. Samples were transferred to local oil mills in solid, vented, food-grade harvest bins or in suitable waist harvest bags. Olive mills were equipped with two or three-phase centrifugal systems (decanters), as olive mills in Messinia operate with both extraction methods (the ratio of two- and three-phase olive mills in Messinia is approximately 50:50). Olive fruits were processed within 24 h, according to the relevant EC Regulation for Kalamata PDO olive oil and the same post-harvest conditions were maintained in all cases. In detail, the leaves were removed from the olive fruits, washed and then sent to the crusher. Malaxation was carried out at low temperatures (27–28 °C) for 30 min according to the above-mentioned regulation. The obtained olive paste was horizontally centrifuged (decanted) (three- or two-phase system) and the resulting olive oil was finally centrifuged. Olive oil samples were stored directly in 1 L air-tight dark-green glass bottles at 4 °C until further analysis. Quality parameters were analyzed in triplicate, while all the other examined chemical parameters were determined in duplicate.

2.3. Determination of the Physicochemical Quality Parameters

Free acidity, peroxide value and spectroscopic indices (K_{232} and K_{268}) were carried out, following the analytical methods described in Regulation EEC/2568/91 of the European Commission and later amendments [23]. Free acidity was expressed as the percentage of oleic acid and peroxide value was given as milliequivalents of active oxygen per kilogram of oil (meq O_2 kg^{-1}). K_{232} and K_{268} extinction coefficients were calculated from absorption at 232 and 268 nm respectively. Spectrophotometric examination in the ultraviolet provides information on the olive oil quality, its state of preservation and changes brought about by technological processes (due to the presence of conjugated diene and triene systems resulting mainly from oxidation processes). These absorptions are expressed as specific extinctions E (the extinction of 1% w/v solution of the oil in isooctane, in a 10 mm cell) conventionally indicated by K 'extinction coefficient'. Free acidity (FA), peroxide value (PV), K_{232} and K_{268} were immediately determined for each sample in order to avoid any kind of olive oil deterioration. Solvents used were purchased from Sigma (St. Louis, MO, USA).

2.4. Determination of Sterols and Triterpene Dialcohols

The individual sterols, total sterols and triterpene dialcohols were determined according to the method adopted by EEC/2568/91 regulation, Annexes V with later amendments [23]. The oil sample, with added α-cholestanol (Sigma, St. Louis, MO, USA), as an internal standard, was saponified with potassium hydroxide in ethanolic solution and the unsaponifiable matter was extracted with diethyl ether. The sterol and triterpene dialcohol fractions were separated from the unsaponifiable matter by thin-layer chromatography on a basic silica gel plate (Fluka, Buchs, Switzerland). The fractions recovered from the silica gel were transformed into trimethylsilyl ethers (TMSE) by the addition of pyridine-hexamethyldisilizane-tri-methylchlorosilane (9:3:1, *v/v/v*) (Supelco, Bellefonte, PA, USA). Sterols (%) and triterpene dialcohol contents were determined with a Shimadzu (GC-2010) gas chromatograph equipped with a flame ionization detector (FID), a DB-5 (30 m × 0.32 mm × 0.25 µm) capillary column and an autosampler injector. The operating conditions were as follows: injection temperature 280 °C, column temperature 265 °C, detector temperature 310 °C, splitting ratio (1:50), flow rate 1.4 mL/min and amount of substance injected 1 µL of TMSE solution. The sterols and triterpene dialcohols were eluted in the following order: cholesterol, 24-methylen-cholesterol, campesterol, campestanol, stigmasterol, Δ7-campesterol, Δ5,23-stigmastadienol, clerosterol, β-sitosterol, sitostanol, Δ5-avenasterol, Δ5,24-stigmastadienol, Δ7-stigmastenol, Δ7-avenasterol, erythrodiol and uvaol (calculated as total erythrodiol). Individual peaks were identified on the basis of their relative retention times with respect to the internal standard. The sum of Δ5,23-stigmastadienol, clerosterol, β-sitosterol, sitostanol, Δ5-avenasterol, and Δ5,24-stigmastadienol represents apparent b-sitosterol. Mean values of duplicate experiments in each sample were used for further statistical analysis.

2.5. Determination of Fatty Acid Composition

The fatty acid profile was determined according to the official method of the Regulation EEC/2568/91, Annex IV with amendments [23]. The fatty acid methyl esters (FAME) were obtained by cold alkaline transesterification with methanolic potassium hydroxide solution and extracted with n-heptane. FAME were analyzed on a model GC-2010 Shimadzu chromatograph, equipped with an BPX-70, (60 m × 0.25 mm × 0.25 µm), capillary column and a flame ionization detector (FID). The carrier gas was helium, with a flow of 1.5 mL/min. The temperatures of the injector and detector were set at 250 and 260 °C respectively and the oven temperature was increased gradually from 165 to 225 °C in 35 min. The injection volume was 1 µL. Quantification was achieved using a FAME standard mixture purchased from Sigma (St. Louis, MO, USA). The results were expressed as a percentage of individual fatty acids. Analytical-grade methanol, heptane, and potassium hydroxide were purchased from Sigma (St. Louis, MO, USA).

2.6. Determination of Wax Content

The wax content of olive oil samples was determined according to the Regulation EEC/2568/91, Annex IV with later amendments [23]. A suitable amount of internal standard (lauryl arachidate) was added to 0.5 g of olive oil sample and then fractionized by chromatography on a hydrated silica gel column. The chromatographic elution was carried out with a mixture of n-hexane/diethyl ether, keeping a rate of flow of approximately 15 drops every 10 s. The subsequent fraction was completely dried and finally resolved in 2 mL of n-hexane. Waxes were analyzed on a model GC-2010 Shimadzu chromatograph equipped with an on-column injector, a flame ionization detector and a MEGA-5 HD (10 m × 0.32 × 0.10 mm) capillary column. The operating conditions were as follows: detector temperature 370 °C; the column temperature was increased from 80 to 160 °C at 40 °C/min and up to 340 °C at 5 °C/min for 7 min; the amount of substance injected was 1 µL of the n-hexane solution. The identification of the peaks was based on retention time by comparison with wax mixtures of known retention times analyzed under the same conditions.

2.7. Statistical Analysis

Results were expressed as the mean values ± standard deviation (SD). Data were processed with MINITAB 18 software. Thus, it is possible to extract the minimum and the maximum value of the sample, mean, and standard deviation (SD). Differences between means were tested for statistical significance using analysis of variance (ANOVA). The statistical significance level was set at $p < 0.05$.

Moreover, Principal Component Analysis (PCA) was applied to study the relations between the extraction method (two- or three-phase decanter) on the examined chemical properties.

3. Results and Discussion

3.1. Qualitative Parameter of Messinian Olive Oils, Greece

As shown in Figure 2, of the 71 virgin olive oils analyzed, all are classified as extra virgin olive oil (EVOO), as far as the qualitative indices are concerned, according to the European Regulation (EEC) 2568/91 as amended, with the exception of two samples in total, which were not within the accepted acidity value of 0.80% and excluded for further analysis. The content of free acids is an important quality factor, extensively used as the major criterion for the classification of olive oil at various commercial grades.

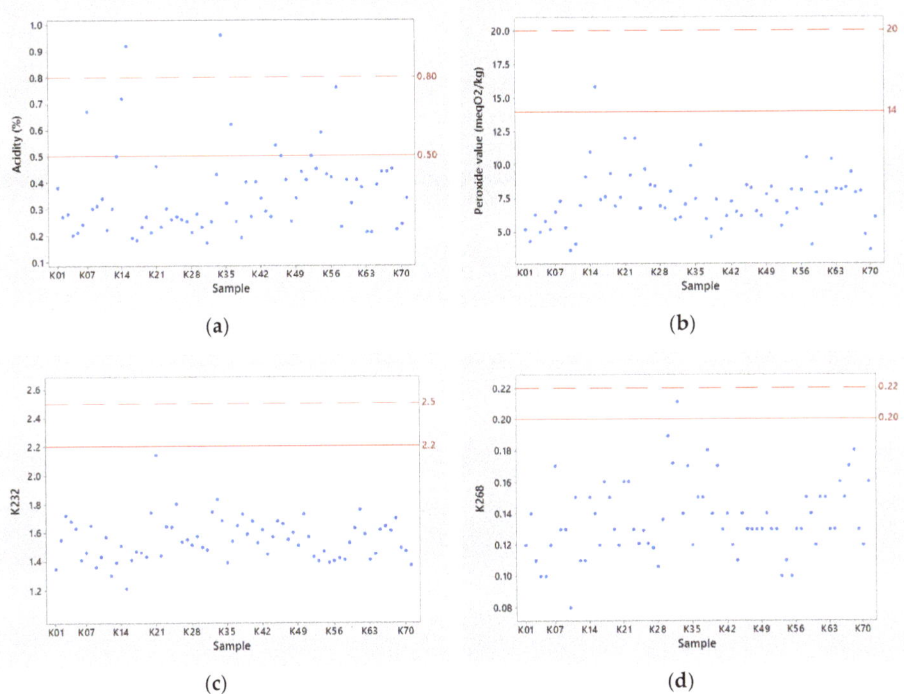

Figure 2. (**a**) Scatter plots visualizing the qualitative parameters (**a**) acidity, (**b**) peroxide value (**c**) K_{232} extinction coefficient and (**d**) K_{268} extinction coefficient, respectively, of the 71 examined Messinian olive oil samples numbered as K1–K71, $N = 71$. Dotted line: limits according to EEC/2568/91 for extra virgin olive oil (EVOO) category; straight line: limits according to Council Regulation (EC) 510/2006 for Kalamata Protected Designation of Origin (PDO) olive oil [20,21,23].

According to the relevant Commission Regulation for Kalamata PDO olive oil, stricter quality specifications compared to EU Regulation 2568/91 have been laid down. [20,21]. As shown in Figure 2, a high percentage of the examined Messinian olive oil samples (88.73%) did not exceed the threshold of 0.50% in acidity, which is defined as the upper limit for Kalamata PDO olive oils. The mean acidity value was 0.34% and ranged from 0.17 to 0.76 (Table 1). In addition, peroxide value and spectrophotometric analysis, crucial indices of olive oil oxidation, were within and quite below the upper limit established by EC Regulation for the EVOO category as presented in Figure 2. In particular, peroxide value for the tested samples ranged from 3.64 to 11.96 meq. O_2 kg^{-1}, with a mean value at 7.24 meq. O_2 kg^{-1} (Table 1). Likewise, K_{232} and K_{268} values had a mean value of 1.55 and 0.13, respectively, with only one sample surpassing the Kalamata PDO limit for K_{268} value.

Table 1. Qualitative parameters of Messinian olive oils.

Parameter	Mean ± SD	Min–Max	EEC Limit for the EVOO Category	PDO Limit
Free acidity (%)	0.34 ± 0.13	0.17–0.76	≤0.80	≤0.50
Peroxide value (meqO_2 kg^{-1})	7.24 ± 1.88	3.64–11.96	≤20	≤14
K_{232}	1.55 ± 0.14	1.33–2.14	≤2.50	≤2.20
K_{268}	0.13 ± 0.01	0.08–0.21	≤0.22	≤0.20

Results are expressed as the means ± standard deviation (SD). N = 69. EEC = European Commission, EVOO = Extra Virgin Olive Oil, PDO = Protected Designation of Origin.

It should be noted that as peroxide value is a quality indicator of the primary products of auto-oxidation (hydroperoxides) of an olive oil, poor post-extraction conditions (e.g., inappropriate storage and packaging) may result to a fast increase in peroxide value above the defined limit of 14 meq $O_2 \cdot$kg^{-1}, excluding olive oils from the PDO labeling.

In general, the above observations depict the highest quality of Messinian olive oil production, one of the most important olive-growing regions in Greece, but most importantly highlight how crucial it is to retain those qualitative characteristics, especially with the recent approval of the European commission to expand "Kalamata PDO olive oil" in the whole regional unit of Messinia (21).

3.2. Analysis of Sterolic Profile and Triterpenic Dialcohol Content of Messinian Olive Oils, Greece

Phytosterols are important components of the unsaponifiable fraction of olive oil beneficial for the human health and nutrition. Sterol composition and content are broadly used for the control of olive oil authenticity and adulteration. Sterol content varies between 1000 and 3000 mg/kg depending the botanical variety, olive ripening, storage conditions and geographical origin [26–31]. Numerous studies have shown that each variety has a characteristic sterol "fingerprint". Therefore, those minor components can be considered as an important and useful tool for detecting oil adulteration and/or classifying virgin olive oils in accordance with their variety [32–35]. The influence of geographical origin on the sterol composition of virgin olive oil has been evaluated by various authors, pointing out the great potential of different analytical techniques followed by chemometric tools for this purpose [36–38].

Although several studies have been conducted for cv Koroneiki in other regions of Greece, mainly in Crete [39–43], very little information is available in the literature regarding the sterolic profile of cv Koroneiki in Peloponnese generally and more precisely in the Messinia region.

In the present study, we evaluated the sterolic composition of the examined Messinian olive oils. Table 2 lists the mean values expressed as percentages of the total sterols and their standard deviations of the main sterols present in the olive oil sampled. The main sterols detected were β–sitosterol, Δ5-avenasterol and campesterol, with mean values of 80.73%, 12.28% and 3.71%, respectively. The first two represent over 90% of the total sterol content, with β-sitosterol being the most abundant phytosterol (over 80% of the total sterol content). The calculated parameter, the apparent β-sitosterol, falls within the established regulatory limits, with a mean value of 94.63%. Finally, the cholesterol and Δ7-stigmastenol values were low and quite below the limits set by EU regulation (0.5%), with a mean value of 0.11% and 0.19% of total sterols, respectively (Table 2).

In contrast, several major deviations were observed in the case of the sterolic profile for the Messinian olive oils. Most importantly, 43.5% of the examined olive oil samples did not surpass the required limit of 1000 mg/kg in total sterol concentration according to the EEC Regulation 2568/91. In addition, as illustrated in Figure 3, the regulated limit for Kalamata PDO olive oil is established at 1100 mg/kg. As a result, a really high percentage (66.7%) of the examined samples was below the established PDO limit. The mean total sterols content was 1033 mg/kg and ranged from 744 to

1283 mg/kg. A similar case was observed in campesterol, where a total of 21.7% of the examined samples exceeded the legal maximum of 4%, with a mean value of 3.71% and ranged from 2.78 to 4.70%. A trend of higher campesterol has also been reported for cv Koroneiki, as well as for other cultivars cultivated in different countries [44,45], whereas the total sterol concentration of the most studied Spanish and Italian cultivars is always within the minimum limit of 1000 mg/kg [46–50].

Table 2. Sterolic profile and triterpene diols determined in Messinian olive oil, Greece.

Sterols and Triterpene Diols	Mean ± SD	EEC Limit	PDO Limit
Cholesterol (%)	0.11 ± 0.03	≤0.5	≤0.5
24-methylene-cholesterol%	0.32 ± 0.09		
Campesterol%	3.71 ± 0.38	≤4.0	≤4.0
Campestanol%	0.05 ± 0.03	<campesterol	<campesterol
Stigmasterol%	0.74 ± 0.19		
Chlerosterol%	0.85 ± 0.07		
β-Sitosterol%	80.73 ± 3.73		
Sitostanol%	0.37 ± 0.30		
Δ-5-avenasterol%	12.28 ± 3.96		
Δ-5,24-stigm/dienol%	0.29± 0.10		
Δ-7-stigmastenol%	0.19 ± 0.09	≤0.5	≤0.5
Δ-7-avenasterol%	0.28 ± 0.11		
Apparent b-Sitosterol%	94.63 ± 1.07	≥93.0	≥93.0
Total erythrodiol%	2.85 ± 1.25	≤4.5	≤4.5
Total sterols (mg/kg)	1033.3 ± 150.1	≥1000	>1100

Results are expressed as the means ± standard deviation (SD). N = 69.

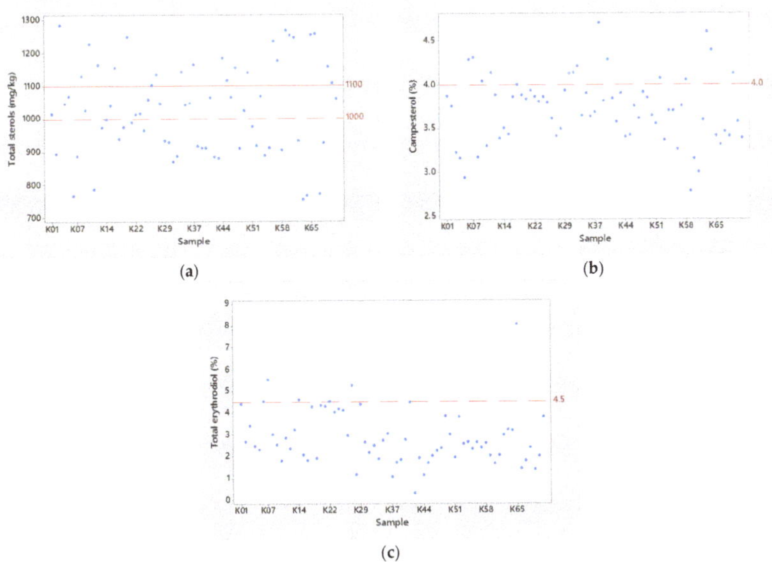

Figure 3. (**a**) Scatter plots visualizing the chemical parameters: (**a**) total sterols (43.5% of the examined olive oil samples did not surpass the EEC limit of 1000 mg/kg and 66.3% of the examined olive oil samples did not surpass the PDO limit of 1100 mg/kg in total sterols); (**b**) campesterol (21.7% of the examined olive oil samples exceeded the legal maximum of 4%); (**c**) total erythrodiol (8.06% of the examined olive oil samples exceeded the upper set limit of 4.5%). Note. Dotted line: limits according to EEC/2568/91 for the EVOO category; straight line: limits according to Council Regulation (EC) 510/2006 for Kalamata PDO olive oil [20,21,23].

Although no information exists in the literature regarding Kalamata PDO olive oils, results show that cv Koroneiki in the Messinian region shows a clear tendency of low concentrations of total sterols and high concentrations of campesterol. Low mean values on total sterol concentration for cv Koroneiki were reported earlier in Crete, in 2001, by Stefanoudaki et al., who studied the effect of drought stress on olive oil characteristics, without giving emphasis on the mentioned tendency [41].

In general, such problems (fluctuations from EU regulations) could inevitably raise questions regarding the authenticity of Kalamata PDO extra virgin olive oils in the olive oil sector, and so they certainly require further investigation.

It is known that total erythrodiol levels are high in solvent-extracted oils, indicating adulteration with olive-pomace oil [51]. The mean total erythrodiol content was 2.85% (Table 2). However, a small but noteworthy percentage of 8.06% of the examined samples exceeded the upper set limit of 4.5% as shown in Figure 3. A possible assumption may be the inappropriate higher degree of olive crushing during the extraction process, leading to an increase in erythordiol levels from the olive's exocarp.

Finally, almost no significant differences were observed in the sterol composition and triterpene dialcohols using the two industrial decanters ($p > 0.05$) (please see Supplementary data). The amount of water added during oil extraction does not affect their levels due to their lipophilic nature and because they are sparingly soluble in water. This is in agreement with previous reported data for cv Koroneiki among other cultivars [40,52].

3.3. Fatty Acid Composition of Messinian Olive Oil, Greece

A crucial parameter for the quality and characterization of olive oil is the fatty acid composition [53]. In the present study, thirteen fatty acids were identified. As shown in Table 3, the variability of fatty acid composition was within the normal range expected for the EVOO category in all the examined samples. The mean values for the major fatty acids were 76.70% for oleic acid (C18:1), 12.02% for palmitic acid (C16:0), 6.09% for linoleic acid (C18:2), 2.53% for stearic acid (C18:0), and 0.92% for palmitoleic acid (C16:1). The percentage of the monounsaturated oleic acid ranged from 70.67% to 81.40% and depicts the beneficial health impact of Messinian olive oils and the competitive profile of Kalamata PDO olive oils to the olive oil market. Palmitic acid, the second most abundant fatty acid ranged from 9.54% to 13.56%, the poly-unsaturated linoleic acid ranged between 4.2% and 12.01%, stearic ranged from 1.98% to 3.12% and palmitoleic acid ranged from 0.64% to 1.43%.

Table 3. Percentage composition (%) of major fatty acids in Messinian olive oils and influence of the extraction method on the fatty acid profile.

Fatty Acid (%)	Mean ± SD	Min–Max	EEC Limit	PDO Limit	Two-Phase Mean ± SD	Three-Phase Mean ± SD	Difference *p*-Value
Myristic C14:0	0.01 ± 0.00	0.00–0.02	≤0.03		0.01 ± 0.00	0.01 ± 0.00	n.s
Palmitic C16:0	12.02 ± 0.74	9.54–13.56	7.50–20.00	10.0–15.0	11.96 ± 0.81	12.11 ± 0.61	n.s
Palmitoleic C16:1	0.92 ± 0.13	0.64–1.43	0.30–3.50	0.6–1.2	0.93 ± 0.13	0.89 ± 0.08	n.s
Heptadecanoic C17:0	0.05 ± 0.02	0.03–0.15	≤0.40		0.04 ± 0.01	0.06 ± 0.03	0.003
Heptadecenoic C17:1	0.08 ± 0.04	0.06–0.24	≤0.60		0.07 ± 0.01	0.10 ± 0.05	0.016
Stearic C18:0	2.53 ± 0.19	1.98–3.12	0.50–5.00	2.0–4.0	2.44 ± 0.16	2.67 ± 0.16	0.00
Oleic C18:1	76.70 ± 1.96	70.67–81.40	55.00–83.00	70–80	76.92 ± 2.19	76.36 ± 1.52	n.s
Linoleic C18:2	6.09 ± 1.60	4.20–12.01	2.50–21.00	4.0–11.0	6.05 ± 1.91	6.14 ± 0.95	n.s
Linolenic C18:3	0.68 ± 0.07	0.51–0.86	≤1.00		0.66 ± 0.08	0.69 ± 0.06	n.s
Arachidic C20:0	0.44 ± 0.03	0.33–0.50	≤0.60		0.43 ± 0.04	0.45 ± 0.02	0.012
Eicosenoic C20:1	0.31 ± 0.02	0.27–0.35	≤0.50		0.31 ± 0.02	0.30 ± 0.02	n.s
Behenic C22:0	0.14 ± 0.01	0.09–0.17	≤0.20		0.14 ± 0.01	0.14 ± 0.01	n.s
Lignoceric C24:0	0.05 ± 0.00	0.034–0.08	≤0.20		0.05 ± 0.00	0.05 ± 0.00	n.s

Results are expressed as the mean ± standard deviation (SD). n.s = non-significant. Differences between means of two-phase vs. three-phase centrifugal systems were tested for statistical significance using analysis of variance (ANOVA). The statistical significance level was set at $p < 0.05$.

Despite the fact that all samples met the standards for the EVOO category as mentioned above, according to the EEC regulation 2568/91, we should stress that quite narrower limits exist for specific fatty acids such as palmitoleic, stearic, oleic and linoleic acid, according to the relevant EU regulation for Kalamata PDO olive oil as shown in Table 3 [20,21]. This led to fluctuations in a number of cases (10/69) from the legal PDO limits. Hence, further investigation should be carried out.

Finally, our results showed that there were only minor changes in fatty acid composition when different decanters were used. In particular, as shown in Table 3, the only significant differences were observed in C17:0, C17:1, C18:0 and C20:0 which are fatty acids of lower proportions in the total fatty acid percentage. This is in agreement with several other studies, where it is reported that there is very little effect of the extraction method on fatty acid composition [40,54,55].

3.4. Analysis of Wax Content of Messinian Olive Oil, Greece

Waxes are important constituents of olive oil used in order to distinguish olive oil obtained by pressing and that obtained by extraction (olive-residue oil) [56,57]. Waxes are present on the external fruit wax cuticle in olives so as to protect the fruit from transpiration and insect damage [57]. In dry hot weather, plants produce more waxes in order to control the rate of transpiration so that reduction of water loss is achieved. As a result, high temperature increases the wax production as a mechanism of fruit defense from environmental factors (climate) [58]. Generally, it has been found that wax compositions are influenced mainly by cultivar, harvest year and malaxation conditions [59–61].

In the present study, six wax esters were detected in the GC chromatogram: C36, C38, C40, C42, C44 and C46. Due to the latest amendment of the EEC regulation [23], the sum of wax esters was classified in two groups. Since only C42, C44 and C46 are now included in the most recent EU regulation for extra virgin olive oil, they were grouped together as Wax Esters (WEs 42–46). Since the sum of C40, C42, C44 and C46 was previously calculated, they were also grouped together as Total Wax Esters (TWEs 40–46). As shown in Table 4, the mean value of WEs 42–46 of the examined olive oil samples was 28.38 mg/kg and ranged from 16.89 to 58.33 mg/kg, well short of the upper legal limit of 150 mg/kg. Respectively, the TWEs 40–46 had a mean value of 67.20 mg/kg and ranged between 42.84 to 140.31 mg/kg. Although there is no previous study on wax determination for Kalamata PDO olive oils, according to our knowledge, the obtained results were similar to the wax content of olive oils extracted in other hot climates such as those in southern Italy for cv Carolea and in Australia for cv Koroneiki [44.46]. Finally, we found no differences in wax content caused by extraction ($p > 0.05$) (please see Supplementary data).

Table 4. Wax esters of Kalamata PDO olive oils produced in Messinia (southwest of Peloponesse).

Parameter (mg/kg)	Mean ± SD	Min–Max	EEC Limit	PDO Limit
Wax Esters C40–C46 (WEs)	67.20 ± 18.88	42.84–140.31	≤250	≤250
Wax Esters C42–C46 (TWEs)	28.38 ± 9.62	16.89–58.33	≤150	≤150

Results are expressed as the means ± standard deviation (SD).

Using chemometric analysis, we verified that the extraction method (two- or three-phase decanter) causes non or minor changes on the examined chemical characteristics.

The score plot of Principal Component Analysis (PCA) is used to assess the data structure and detect clusters, outliers, and trends. As shown in Figure 4 groupings of data on the plot based on two-phase and three-phase decanters in Messinian olive oil samples showed that the points are randomly distributed around zero. As a result, no correlations for both two-phase and three-phase samples were presented. Therefore, the extraction method (two- or three-phase decanter) has verified no changes to the chemical parameters examined.

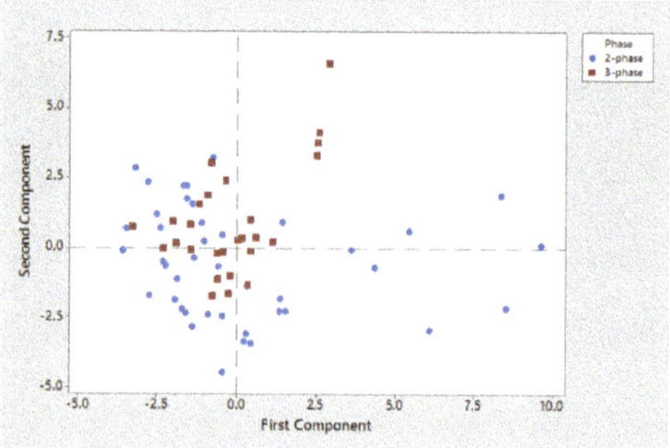

Figure 4. Score plot of Principal Component Analysis (PCA) for two-phase and three-phase decanters in Messinian olive oil.

4. Conclusions

The evaluation of extra virgin olive oils produced in the Messinia region, southwest of Peloponnese, denoted some challengeable characteristics. On the one hand, the results depict the high qualitative profile of Messinian olive oils, which is in agreement with similar studies examining cv Koroneiki from different geographical regions of Greece. On the other hand, major fluctuations were observed from the established EU regulatory limits. Most importantly, results show that Messinian extra virgin olive oils show low concentration of total sterols, with 66.7% of the examined samples being below the regulated set limits for Kalamata PDO status. Although no information exists in the literature regarding Kalamata PDO olive oils, as mentioned previously, analysis of VOOs from cv Koroneiki in completely different geographical regions such as Crete and Australia, has also shown a tendency of low total sterol concentration. Thus, low mean value in total sterols may clearly depict a "special characteristics" for Koroneiki cultivar, yet completely opposed to the existing standard limits of Kalamata PDO status. In contrast, total sterol concentration of the most studied Spanish and Italian cultivars is always quite above the limit of 1000 mg/kg.

In addition, our results show that olive oil samples of cv Koroneiki in the Messinia region present a high concentration of campesterol, with a total of 21.7%, exceeding the legal maximum of 4.0% and a light tendency of high total erythrodiol content. A trend of higher campesterol has been reported for cv Koroneiki, cv Barnea and cv Cornicabra cultivated in other geographical regions such as in Australia and Spain. Furthermore, although the fatty acid composition of the examined samples was within the range for the EVOO category, the extremely narrow established PDO limits in specific fatty acid composition may result in further fluctuations excluding Messinian olive oils from PDO certification. As far as wax content is concerned, although no information exists in the literature for Messinian olive oils for comparison, the obtained results are within the regulatory EU limits. Finally, in accordance with previous reported data for cv Koroneiki in Greece, the extraction method (two- or three-phase decanter) caused non or minor changes on the examined chemical characteristic.

In general, as PDO-certified products are a crucial strategic tool to enhance rural economy and development, through the added value of the PDO trademark, in terms of the higher price such products can enjoy, the above-mentioned deviations could inevitably lead to a controversy regarding the authenticity of Kalamata PDO extra virgin olive oils in the olive oil sector and consequently result in diminishing its reputation.

As this work is the first systematic attempt focusing on the evaluation of "Kalamata PDO olive oil" characteristics, further in depth research, with a higher number of samples and more crop years, is under way. The continued study of Messinian olive oils with the addition of more examined parameters (e.g., sensory analysis) will provide adequate datasets and allow supporting the improvement of the current EU regulation through the update and the re-adjustment of the established limits for Kalamata PDO status.

Supplementary Materials: The following are available online at http://www.mdpi.com/2304-8158/8/12/610/s1, Table S1: Influence of the extraction method on the sterolic profile, Table S2: Influence of the extraction method on wax esters content.

Author Contributions: V.S. designed and performed the experiments; P.T. performed statistical analysis; T.V. and V.S. wrote, edited and reviewed the paper.

Funding: This research project was funded under the Project 'Research and Technology Development Innovation Projects'-AgroETAK, MIS 453350, in the framework of the Operational Program 'Human Resources Development'. It is co-funded by the European Social Fund through the National Strategic Reference Framework (Research Funding Program 2007–2013) coordinated by the Hellenic Agricultural Organization–DEMETER.

Acknowledgments: V.S. would like to thank A.S. Vekiari and the Dept. of Olive and Horticultural crops, Kalamata, Greece, for the support provided. All the experiments were conducted in the accredited laboratory of the Dept. of Olive and Horticultural crops, Kalamata, Greece, according to ISO 17025:2008.

Conflicts of Interest: The authors declare no conflict of interest.

References

1. Preddy, V.; Watson, R. *Olives and Olive Oil in Health and Disease Prevention*, 1st ed.; Academic Press: Oxford, UK, 2010; pp. 1100–1220.
2. Barbouti, A.; Briasoulis, E.; Galaris, D. Protective effects of olive oil components against hydrogen peroxide-induced DNA damage. In *Olives and Olive Oil in Health and Disease Prevention*; Preddy, V., Watson, R., Eds.; Academic Press: Oxford, UK, 2010; pp. 1103–1109.
3. Ragusa, A.; Centonze, C.; Grasso, M.E.; Latronico, M.F.; Mastrangelo, P.F.; Fanizzi, F.P.; Maffia, M. Composition and Statistical Analysis of Biophenols in Apulian Italian EVOOs. *Foods* **2017**, *6*, 90. [CrossRef]
4. Borges, T.H.; Serna, A.; López, L.C.; Lara, L.; Nieto, R.; Seiquer, I. Composition and Antioxidant Properties of Spanish Extra Virgin Olive Oil Regarding Cultivar, Harvest Year and Crop Stage. *Antioxidants* **2019**, *8*, 217. [CrossRef]
5. Kalogeropoulos, N.; Tsimidou, M.Z. Antioxidants in Greek Virgin Olive Oils. *Antioxidants* **2014**, *3*, 387–413. [CrossRef]
6. Boskou, D. *Olive Oil: Chemistry and Technology*, 2nd ed.; AOCS Press: Champaign, IL, USA, 2006; pp. 45–93.
7. Prospects for the Olive Oil Sector in Spain, Italy and Greece 2012–2020. Available online: https://ec.europa.eu/agriculture/markets-and-prices/market-briefs_en (accessed on 12 June 2019).
8. International Olive Council Database. Available online: http://www.internationaloliveoil.org/estaticos/view/131-world-olive-oil-figures (accessed on 17 August 2019).
9. Zampounis, V. Olive oil in the World Market. In *Olive Oil: Chemistry and Technology*, 2nd ed.; Boskou, D., Ed.; AOCS Press: Campaign, IL, USA, 2006; pp. 21–40.
10. Olive Oil: Establishing the Greek brand, National Bank of Greece. Available online: https://www.nbg.gr/greek/the-group/press-office/e-spot/reports/Documents/Olive%20Oil_2015.pdf (accessed on 23 May 2019).
11. Kostalenos, G. *Olive Fruit Data, History, Description and Geographical Distribution of Olive Varieties in Greece*, 1st ed.; Kostalenos: Poros Trizinias, Greece, 2011; pp. 1–436. (In Greek)
12. Agiomirgianaki, A.; Petrakis, P.; Dais, P. Influence of harvest year, cultivar and geographical origin on Greek extra virgin olive oils composition: A study by NMR spectroscopy and biometric analysis. *Food Chem.* **2012**, *135*, 2561–2568.
13. Karabagias, I.; Michos, C.; Badeka, A.; Kontakos, S.; Stratis, I.; Kontominas, M.G. Classification of Western Greek virgin olive oils according to geographical origin based on chromatographic, spectroscopic, conventional and chemometric analyses. *Food Res. Int.* **2013**, *54*, 1950–1958. [CrossRef]
14. Kandylis, P.; Vekiari, A.S.; Kanellaki, M.; Grati Kamoun, N.; Msallem, M.; Kourkoutas, Y. Comparative study of extra virgin olive oil flavor profile of Koroneiki variety (Olea europaea var. Microcarpa alba) cultivated in Greece and Tunisia during one period of harvesting. *LWT J. Food Sci. Technol.* **2011**, *44*, 1333–1341. [CrossRef]
15. Longobardi, F.; Ventrella, A.; Casielloa, G.; Sacco, D.; Tasioula-Margari, A.K.; Kiritsakis, A.K.; Kontominas, M.G. Characterisation of the geographical origin of Western Greek virgin olive oils based on instrumental and multivariate statistical analysis. *Food Chem.* **2012**, *133*, 1169–1175. [CrossRef]

16. Pouliarekou, E.; Badeka, A.; Tasioula-Margaria, M.; Kontakos, S.; Longobardic, F.; Kontominas, M.G. Characterization and classification of Western Greek olive oils according to cultivar and geographical origin based on volatile compounds. *J. Chromatogr. A* **2011**, *1218*, 7534–7542. [CrossRef]

17. Kotsiou, K.; Tasioula-Margari, M. Monitoring the phenolic compounds of Greek extra-virgin olive oils during storage. *Food Chem.* **2016**, *200*, 255–262. [CrossRef]

18. Council Regulation (EC). No. 510/2006 of 20 March 2006 on the protection of geographical indications and designations of origin for agricultural products and foodstuffs. *Off. J. Eur. Union* **2006**, *L93*, 12–25.

19. Commission Regulation (EC). No. 1898/2006 of 14 December 2006 laying down detailed rules of implementation of Council Regulation (EC) No. 510/2006 on the protection of geographical indications and designations of origin for agricultural products and foodstuffs. *Off. J. Eur. Union* **2006**, *L369*, 1–23.

20. Official website of the European Union. Available online: https://op.europa.eu/en/publication-detail/-/publication/a974b890-424a-11e5-9f5a-01aa75ed71a1/language-en (accessed on 12 June 2019).

21. EUR-Lex. Access to European Law. Available online: https://eur-lex.europa.eu/legal-content/EN/TXT/?qid=1574338279830&uri=CELEX:52012XC0626(03) (accessed on 12 June 2019).

22. Messinia Perfecture. Available online: https://el.wikipedia.org/wiki/%CE%9D%CE%BF%CE%BC%CF%8C%CF%82_%CE%9C%CE%B5%CF%83%CF%83%CE%B7%CE%BD%CE%AF%CE%B1%CF%82 (accessed on 12 June 2019).

23. Commission Regulation (EEC). No. 2568/91 of 14 July 1991 on the characteristics of olive oil and olive-residue oil and on the relevant methods of analysis. *Off. J.* **1991**, *L208*, 1–8.

24. Antonini, E.; Farina, A.; Leone, A.; Mazzara, E.; Urbani, S.; Selvaggini, R.; Servili, M.; Ninfali, P. Phenolic compounds and quality parameters of family farming versus protected designation of origin (PDO) extra-virgin olive oils. *J. Food Compos. Anal.* **2016**, *43*, 75–81. [CrossRef]

25. Egea, P.; Pérez y Pérez, L. Sustainability and multifunctionality of protected designations of origin of olive oil in Spain. *Land Use Policy* **2016**, *58*, 264–275. [CrossRef]

26. Moreau, R.A.; Norton, R.A.; Hicks, K.B. Phytosterols and phytostanols lower cholesterol. *Inform* **1999**, *10*, 572–577.

27. Morchio, G.; De Anreis, R.; Fedeli, E. Investigations on total sterols content in the olive oil and their variation during the refining process. *Riv. Ital. Delle Sostanze Grasse* **1987**, *64*, 185–192.

28. Boskou, D.; Tsimidou, M.; Blekas, D. Olive oil composition. In *Olive Oil Chemistry and Technology*; Boskou, D., Ed.; AOCS Press: Champaign, IL, USA, 2006; pp. 41–72.

29. López-Miranda, J.; Pérez-Jiménez, F.; Ros, E.; De Caterina, R.; Badimón, L.; Covas, M.I.; Escrich, E.; Ordovás, J.M.; Soriguer, F.; Abiá, R.; et al. Olive oil and health: Summary of the II international conference on olive oil and health consensus report, Jaén and Córdoba (Spain). *Nutr. Metab. Cardiovasc. Dis.* **2010**, *20*, 284–294. [CrossRef]

30. Matos, L.C.; Cunha, S.C.; Amaral, J.S.; Pereira, J.A.; Andrade, P.B.; Seabra, R.M.; Andrade, P.B.; Seabra, R.M. Chemometric characterization of three varietal olive oils (Cvs. Cobrancosa, Madural and Verdeal Transmontana) extracted from olives with different maturation indices. *Food Chem.* **2007**, *102*, 406–414. [CrossRef]

31. Aparicio, R.; Luna, G. Characterization of monovarietal virgin olive oil. *J. Lipid Sci. Technol.* **2002**, *104*, 614–627. [CrossRef]

32. Lukic, M.; Lukic, I.; Krapac, M.; Sladonja, B.; Pilizota, V. Sterols and triterpene diols in olive oil as indicators of variety and degree of ripening. *Food Chem.* **2013**, *136*, 251–258. [CrossRef]

33. Galeano, T.; Durán, I.; Sánchez, J.; Alexandre, M.F. Characterization of virgin olive oil according to its triglycerides and sterols composition by chemometric methods. *Food Control* **2005**, *16*, 339–347. [CrossRef]

34. Mohameda, M.B.; Rocchettid, G.; Montesanoe, D.; Ali, S.B.; Guasmib, F.; Grati-Kamouna, N.; Lucinif, L. Discrimination of Tunisian and Italian extra-virgin olive oils according to their phenolic and sterolic fingerprints. *Food Res. Int.* **2018**, *106*, 920–927. [CrossRef] [PubMed]

35. Aparicio, R.; García-González, D.L. Olive oil characterization and traceability. In *Handbook of Olive Oil: Analysis and Properties*; Aparicio, R., García-González, D.L., Eds.; Springer: New York, NY, USA, 2013; Volume 12, pp. 431–478.

36. Temine, S.B.; Manai, H.; Methenni, K.; Baccouri, B.; Abaza, L.; Daoud, D.; Zarrouk, M. Sterolic composition of Chétoui virgin olive oil: Influence of geographical origin. *Food Chem.* **2008**, *110*, 368–374. [CrossRef] [PubMed]

37. Giacalone, R.; Giuliano, S.; Gulotta, E.; Monfreda, M.; Presti, G. Origin assessment of EV olive oils by esterified sterols analysis. *Food Chem.* **2015**, *188*, 279–285. [CrossRef] [PubMed]
38. Bagur-González, M.G.; Pérez-Castano, D.; Sánchez-Vinas, M.; Gázquez-Evangelista, D. Using the liquid-chromatography-fingerprint of sterols fraction to discriminate virgin olive from other edible oils. *J. Chromatogr.* **2015**, *138*, 64–70. [CrossRef]
39. Koutsaftakis, A.; Kotsifaki, F.; Stefanoudaki, E. Effect of extraction system, stage of ripeness and kneading temperature on the sterol composition of virgin olive oils. *J. Am. Oil Chem. Soc.* **1999**, *76*, 1477–1481. [CrossRef]
40. Stefanoudaki, E.; Koutsaftakis, A.; Harwood, J.L. Influence of malaxation conditions on characteristic qualities of olive oils. *Food Chem.* **2011**, *127*, 1481–1486. [CrossRef]
41. Stefanoudaki, E.; Chartzoulakis, K.; Koutsaftakis, A.; Kotsifaki, F. Effect of drought stress on qualitative characteristics of olive oil of cv Koroneiki. *Grasas y Aceites* **2001**, *52*, 202–206. [CrossRef]
42. Vekiari, S.A.; Oreopoulou, V.; Kourkoutas, Y.; Kamoun, N.; Msallem, M.; Psimouli, V.; Arapoglou, D. Characterization and seasonal variation of the quality of virgin olive oil of the Throumbolia and Koroneiki varieties from Southern Greece. *Grasas y Aceites* **2010**, *61*, 221–231. [CrossRef]
43. Koutsaftakis, A.; Kotsifaki, F.; Stefanoudaki, E. Characterization of Cretan extra-virgin olive oils from Koroneiki variety. Influence of the location of origin on several analytical parameters. *Olivae* **2000**, *81*, 20–25.
44. Mailer, R.J.; Ayton, J.; Graham, K. The Influence of growing region, cultivar and harvest timing on the diversity of Australian olive oil. *J. Am. Oil Chem. Soc.* **2010**, *87*, 877–884. [CrossRef]
45. Rivera del Alamo, R.M.; Fregapane, G.; Aranda, F.; Gomez-Alonso, S.; Salvador, M.D. Sterol and alcohol composition of Cornicabra virgin olive oil: The campesterol content exceeds the upper limit of 4% established by EU regulations. *Food Chem.* **2004**, *84*, 533–537. [CrossRef]
46. Piscopo, A.; De Bruno, A.; Zappia, A.; Ventre, C.; Poiana, M. Characterization of monovarietal olive oils obtained from mills of Calabria region (Southern Italy). *Food Chem.* **2016**, *213*, 313–318. [CrossRef] [PubMed]
47. Martinez Cano, M.; Gordillo, C.M.; Mendoza, M.F.; Vertedor, M.F.; Sanchez Casas, J. The Sterol and Erythrodiol + Uvaol Content of Virgin Olive Oils Produced in Five Olive-Growing Zones of Extremadura (Spain). *J. Am. Oil Chem. Soc.* **2015**, *93*, 227–235. [CrossRef]
48. Giuffre, A.M.; Louadj, L.; Poiana, M.; Macario, A. Sterol composition of oils extracted from olives cultivars of the province of Reggio Calabria (south of Italy). *Riv. Ital. Delle Sostanze Grasse* **2012**, *89*, 177–183.
49. Salvador, M.D.; Aranda, F.; Fregapane, G. Influence of fruit ripening on "Cornicabra" virgin olive oil quality. A study of four successive crop seasons. *Food Chem.* **2001**, *73*, 45–53. [CrossRef]
50. Sanchez Casas, J.; Osorio Bueno, E.; Montano Garcia, A.M.; Martinez Cano, M. Sterol and erythrodiol + uvaol content of virgin olive oils from cultivars of Extramadura (Spain). *Food Chem.* **2004**, *87*, 225–230. [CrossRef]
51. Reina, R.J.; White, K.D.; Jahngen, E.G. Validated method for quantitation and identification of 4,4-desmethylsterols and triterpene diols in plant oils by thin layer chromatography-high resolution gas chromatography-mass spectrometry. *J. AOAC Int.* **1997**, *80*, 1272–1280.
52. Di Giovacchino, L.; Sestili, S.; Di Vincenzo, D. Influence of olive processing on virgin olive oil quality. *Eur. J. Lipid Sci. Technol.* **2002**, *104*, 587–601. [CrossRef]
53. Boskou, D. Olive oil. In *More on Mediterranean Diets*; Simopoulos, A.P., Visioli, F., Eds.; Karger: Basel, Switzerland, 2007; Volume 97, pp. 180–210.
54. Gimeno, E.; Castellote, A.I.; Lamuela-Raventós, R.M.; De la Torre, M.C.; López- Sabater, M.C. The effects of harvest and extraction methods on the antioxidant content (phenolics, a-tocopherol, and b-carotene) in virgin olive oil. *Food Chem.* **2002**, *78*, 207–211. [CrossRef]
55. Vaz-Freirea, L.; Gouveia, J.M.J.; Costa Freitasa, A.M. Analytical characteristics of olive oils produced by two different extraction techniques, in the Portuguese olive variety 'Galega Vulgar'. *Grasas y Aceites* **2008**, *59*, 260–266.
56. Kiritsakis, A.; Christie, W. Analysis of edible oil. In *Handbook of Olive Oil*; Harwood, J., Aparicio, R., Eds.; Springer: Aspen, CO, USA; Gaithersburg, MD, USA, 2000; pp. 129–158.
57. Hamilton, R.J. *Waxes: Chemistry, Molecular Biology and Functions*; The Oily Press: Dundee, UK, 1995; pp. 145–189.
58. Bianchi, G. Plant waxes. In *Waxes: Chemistry, Molecular Biology and Functions*; Hamilton, R.J., Ed.; The Oily Press: Dundee, UK, 1995; pp. 175–222.

59. Giuffre, A.M. Influence of harvest year and cultivar on wax composition of olive oils. *Eur. J. Lipid Sci. Technol.* **2013**, *115*, 549–555. [CrossRef]
60. Giuffre, A.M. Wax Ester Variation in Olive Oils Produced in Calabria (Southern Italy) During Olive Ripening. *J. Am. Oil Chem. Soc.* **2014**, *91*, 1355–1366. [CrossRef]
61. Inarejos-García, A.M.; Gómez-Rico, A.; Salvador, M.D.; Fregapane, G. Influence of malaxation conditions on virgin olive oil yield, overall quality and composition. *Eur. Food Res. Technol.* **2009**, *228*, 671–677. [CrossRef]

Article

Inter-Varietal Diversity of Typical Volatile and Phenolic Profiles of Croatian Extra Virgin Olive Oils as Revealed by GC-IT-MS and UPLC-DAD Analysis

Igor Lukić [1,2,*], Marina Lukić [1], Mirella Žanetić [3], Marin Krapac [1], Sara Godena [1] and Karolina Brkić Bubola [1]

[1] Institute of Agriculture and Tourism, Karla Huguesa 8, HR-52440 Poreč, Croatia; marina@iptpo.hr (M.L.); marin@iptpo.hr (M.K.); sara@iptpo.hr (S.G.); karolina@iptpo.hr (K.B.B.)
[2] Centre of Excellence for Biodiversity and Molecular Plant Breeding, Svetošimunska 25, HR-10000 Zagreb, Croatia
[3] Institute for Adriatic Crops and Karst Reclamation, Put Duilova 11, HR-21000 Split, Croatia; mirella.zanetic@krs.hr
* Correspondence: igor@iptpo.hr; Tel.: +385-52-408-327

Received: 11 October 2019; Accepted: 6 November 2019; Published: 9 November 2019

Abstract: Despite having an interesting native olive gene pool and a rapidly emerging olive oil industry, monovarietal extra virgin olive oils (EVOO) from Croatia are relatively unexplored. To investigate the inter-varietal diversity of typical volatile and phenolic profiles of Croatian EVOO, 93 samples from six olive (*Olea europaea* L.) varieties were subjected to gas chromatography-ion trap mass spectrometry (GC-IT-MS) and ultra-performance liquid chromatography with diode array detection (UPLC-DAD), respectively. Quantitative descriptive sensory analysis was also performed. Analysis of variance extracted many relevant exclusive or partial discriminators between monovarietal EVOOs among the identified volatile compounds and phenols. Successful differentiation model with a 100% correct classification was built by linear discriminant analysis, while the most typical volatiles for each monovarietal EVOO were confirmed by partial least squares discriminant analysis. Diverse typical sensory attributes among the EVOOs were tentatively ascribed to the variations in the composition of volatiles and phenols. It was proven that the approach that comprises GC-IT-MS and UPLC-DAD analysis may provide additional objective information about varietal origin and typicity which successfully complement those obtained by sensory analysis. The approach was characterized as universal in nature, with a significant potential to contribute in strengthening the varietal identities and position on the market of monovarietal and Protected Denomination of Origin (PDO) EVOO.

Keywords: extra virgin olive oil; volatile compounds; phenols; sensory quality; varietal typicity

1. Introduction

Extra virgin olive oil (EVOO) is appreciated among consumers because of its specific flavor and nutritional properties. Due to its economic importance, EVOO is among the most common commodities subject to fraud and mislabeling. The European Union (EU) protects EVOO by the regulation mostly based on analytical and sensory controls [1] which generally succeed in detecting illegal manipulation with EVOO intrinsic properties (adulteration with cheaper refined and/or extraneous oils) and EVOO extrinsic properties (fraudulent misrepresentation of quality category). In the EU, most EVOOs of high economic value are additionally protected by Protected Denomination of Origin (PDO) [2]. Each PDO EVOO is produced according to a set of specific requirements prescribed by the holder of a designation in a specification document, governing aspects such as olive varieties used, cultivation, harvest and processing conditions, physico-chemical parameters, and sensory characteristics.

Many PDO EVOOs are produced from olives of a single variety (monovarietal EVOOs), while the blends also owe a large part of their typicity to the unique olive assortment of a particular region. Nowadays, the information on the label about the varietal origin of EVOO is becoming more and more important and attracting, especially for the market segment of informed consumers interested in healthy, quality products with remarkable diversity and clear identity. Similar as in the case of wine, besides being linked to a given geographical origin and PDO, EVOOs from particular varieties are recognized, appreciated, and demanded on the market because of their specific nutritional and sensory properties. As a consequence, they often reach higher prices, and, given the obvious financial benefits associated with them, are very likely subject to fraud by mislabeling with respect to varietal origin.

As regards EVOO varietal authentication within the process of protection by designation of origin (PDO), controls against counterfeiting include auditing of mandatory documentation and records that prove traceability in production and compliance with the requirements set up in PDO specification. The other part is the assessment of the conformity of EVOO with physico-chemical parameters and sensory characteristics laid down in the specification. The analytical parameters controlled more often (e.g., acidity, peroxide value, measurements in ultraviolet, etc.) do not specifically reflect varietal origin, and the limits that are established, although usually stricter than those prescribed by the official EU regulation [1], are regularly not designed to identify olive variety used. Similar applies for the sensory profiles commonly used to describe monovarietal/PDO EVOO [3] which are not highly discriminative.

The mentioned measures are not sufficient to control varietal origin and avoid fraud. Fraud or false labelling might also be detected or confirmed chemically by analysis of other, minor EVOO compounds. The general strategy that is followed in various research laboratories is the detection of as many as possible EVOO constituents from a larger set of samples and application of multivariate statistical analysis to the analytical data in order to build up classification/prediction models based on varietal origin [4,5]. Many EVOO compounds were found useful for this purpose, including sterols [6], tocopherols [7], fatty acids [6,8], etc. The chemical compounds whose amounts are regulated neither by the official EU regulation nor the PDO specifications, but are certainly the most involved in the typical sensory identity of PDO and monovarietal EVOOs and could serve as differentiators based on such criteria, are volatile aroma compounds and phenols [9–14]. In fact, many successful reports were published which confirmed the utility of these constituents for EVOO varietal differentiation [12,14–20]. Volatile fraction of high quality EVOO, which is responsible for its characteristic so-called *green* and *fruity* flavor, consists mainly of C5 and C6 volatiles (aldehydes, ketones, alcohols, and esters) generated enzymatically in the so-called lipoxygenase (LOX) pathway and other subsequent bioprocesses during olive processing. LOX-derived compounds are accompanied by those from other chemical classes, such as hydrocarbons, terpenes, benzenoids, etc. with mostly unknown or minor sensory relevance [12,14,21–23]. Besides being among the most important contributors to EVOO antioxidant activity, phenols, especially secoiridoids which are the most abundant, are responsible for the characteristic EVOO bitterness and pungency [12–14,24,25]. Olive oil phenols are formed mainly by cleavage of their glycosides by hydrolytic enzymes during olive fruit processing and their concentrations are further affected by oxidative degradation catalyzed by polyphenoloxidases and peroxidases [26,27]. The activity of the mentioned enzymes responsible for the formation of both volatile compounds and phenols is strongly genetically predetermined [28,29], which makes these compounds a logical choice for potential varietal markers in EVOO varietal characterization and differentiation studies.

Croatia is the latest country that joined EU in 2013, and some of the most recently registered PDOs are Croatian [30]. Despite relatively small quantities produced in relation to the leading olive oil producing countries, such as Spain, Italy, Greece etc. [31], EVOOs from Croatia are emerging rapidly on the global market and are much appreciated. For example, Croatian EVOOs are often among those awarded with the highest prizes at relevant international competitions, while Istria, one of the most important olive growing and EVOO producing regions in Croatia, has been represented in the first and leading global EVOO guide *Flos Olei* by the largest number of EVOOs among all the regions for the last four years in a row (2015–2018). The olive plantations in Croatia have high genetic diversity, including many native varieties which concentrate close to their area of origin and show a limited geographical dispersion [32]. For this reason, Croatian EVOOs protected by various PDOs owe a significant part of their typicity to the varietal origin of the olives, which certainly becomes most pronounced in the case of monovarietal EVOO. In spite of that, and despite existing reports on the chemical and sensory characteristics of Croatian monovarietal EVOO [33–40], the potential of Croatian native olive varieties to produce diverse and specific EVOO has not been investigated enough to be adequately exploited in designing more unique and robust PDOs.

The main aim of this study was to investigate the inter-varietal diversity of typical volatile and phenolic profiles of Croatian monovarietal EVOOs by gas chromatography-ion trap mass spectrometry (GC-IT-MS) and ultra-performance liquid chromatography with diode array detection (UPLC-DAD), respectively. The approach was tested for the characterization and differentiation of EVOOs made from native varieties grown in the two most important olive growing regions in Croatia, Istria and Dalmatia, with each monovarietal EVOO represented by a heterogeneous sample group in terms of geographical microlocations, growing conditions, harvest date, olive processing technology, and EVOO finalization and storage parameters. It was expected that the results obtained would be useful for improving the understanding of the origins of the typical sensory characteristics of the investigated Croatian monovarietal EVOOs. However, the main premise was that the instrumental techniques utilized would be effective in tracing robust chemical markers among the investigated compounds despite the aforementioned sample heterogeneity, able to provide complementary information about varietal origin to that obtained by sensory analysis. Besides allowing better quality management and control in production, such findings would contribute strengthening the PDO identities and position on the market of Croatian EVOO.

2. Materials and Methods

2.1. EVOO Samples

For this study, the most economically important and widespread Croatian native olive varieties (*Olea europaea* L.) were considered. Representative monovarietal EVOO samples, made from Buža (19 samples), Istarska bjelica (22 samples), and Rosinjola (8 samples) olive varieties specific for the region of Istria, and Oblica (15 samples) and Lastovka (10 samples) olive varieties specific for the region of Dalmatia, were collected from local producers. In addition, representative monovarietal EVOO samples from a widespread variety Leccino (19 samples) grown in Istria were also collected. Detailed climatological data for Istrian and Dalmatia regions in year 2015 are reported in Table S1. EVOOs were selected to cover the maximum possible variability of each production area, and all the samples from the same variety were produced by different producers. Olive fruit samples were hand-picked at the usual maturity level for each cultivar during the local customary harvest period during October/November 2015. The collected EVOO samples were produced in various private mills using contemporary oil extraction equipment with temperature of malaxation kept below 27 °C. After finalization (clarification and storage), market-ready EVOOs were kept at low temperature in amber dark glass bottles prior to analysis, and analyzed during a period of 3 months.

2.2. Sensory Analysis

Quantitative descriptive analysis of EVOO samples was performed by the Panel for sensory analysis of VOO of the Institute of Agriculture and Tourism in Poreč (Croatia), accredited for VOO sensory analysis according to the EN ISO/IEC 17025:2007 standard and authorized by the Croatian Ministry of Agriculture for official VOO testing from 2012, and recognized in continuation by the IOC from 2014. The panel consisted of eight assessors (4 female, 4 male, average age 39) trained and accredited for VOO sensory analysis according to the International Olive Council (IOC) method adopted by the European Commission Regulation [1]. As well, all the tasters have had long-term involvement in EVOO research and have gained large experience in Croatian monovarietal EVOO sensory analysis. Qualitative (selection of descriptors/attributes by consensus and standardization of vocabulary) and quantitative (intensity of perception) criteria of the tasters were attuned by audibly tasting representative samples of Croatian monovarietal EVOOs through several preliminary training sessions. The panel agreed that the sensory attributes which best describe the investigated monovarietal EVOOs were the same for all varieties and among those commonly perceived in EVOO, but differed with respect to the ratios of their intensities. The panel used a modified profile sheet expanded with particular positive odor and taste attributes, which were quantified using a 10 cm unstructured intensity ordinal rating scale from 0 (no perception) to 10 (the highest intensity). For evaluating general quality attributes, a 10-point overall structured rating scale from 0 (the lowest quality) to 10 (the highest quality) was applied. For overall quality evaluation, VOOs were graded with points from 1 (the lowest quality) to 9 (the highest quality). Before each session, the tasters attuned their criteria with respect to the intensities of the perceived sensory attributes by tasting the same standard reference VOO sample, a blend characterized by all the selected sensory attributes/descriptors. According to the sensory analysis, all the investigated samples were classified as EVOO (no defect, fruitiness > 0).

2.3. Chemical Standards and Standard Solutions

Methanol, water, and n-hexane were of HPLC grade purity (Sigma-Aldrich, St. Louis, MO, USA). Pure chemical standards of volatile compounds and phenols were purchased from AccuStandard Inc. (New Haven, CT, USA), Acros Organics (Geel, Belgium), Alfa Aesar (Haverhill, MA, USA), Cayman Chemical Co. (Ann Arbor, MI, USA), Extrasynthese (Genay, France), Fluka (Buchs, Switzerland), Honeywell International Inc. (Morris Plains, NJ, USA), Merck (Darmstadt, Germany), and Sigma-Aldrich. Standard solutions of volatiles were prepared in refined sunflower oil and that of phenols in pure methanol.

2.4. Analysis of Volatile Compounds by GC-IT-MS

Volatile compounds were isolated using headspace solid-phase microextraction (HS-SPME), according to the modified method proposed by Brkić Bubola, Koprivnjak, Sladonja, Škevin, and Belobrajić [41]. SPME fiber used was divinylbenzene/carboxen/polydimethylsiloxane (DVB/CAR/PDMS), 1 cm length, 50/30 μm film thickness (Supelco, Bellefonte, PA, USA). Four grams of EVOO sample (or a standard solution) were placed in a 10 mL glass vial containing a micro-stirring bar, and sealed. The headspace in the vial was equilibrated at 40 °C for 15 min, and the extraction was carried out at 40 °C for 40 min with stirring at 800 rpm. Thermal desorption of analytes was achieved in the GC injection port in splitless mode at 245 °C for 3 min. Identification and quantification of volatile compounds was performed using a Varian 3900 GC coupled to a Varian Saturn 2100 T ion trap mass spectrometer (IT-MS) (Varian Inc., Harbor City, CA, USA). A capillary column Rtx-WAX (60 m × 0.25 mm i.d. × 0.25 μm film thickness; Restek, Bellefonte, PA, USA) was used. Initial oven temperature was 40 °C, increased to 210 °C at 2 °C/min, increased to 245 °C at 20 °C/min, and kept for 20 min. Injector, transfer line and ion trap temperatures were 245, 180, and 120 °C, respectively. Mass spectra were acquired in EI mode (70 eV) at 1 s/scan, full scan with a range of 30–450 *m/z*. The carrier gas was helium (1.2 mL/min).

Identification was performed by comparing retention times and mass spectra with those of pure standards, and with mass spectra from NIST05 library. Identification by comparison with mass spectra was considered satisfactory if spectra reverse match numbers (RM) higher than 800 were obtained. If in a particular sample the mass spectra were not clear (RM < 800), identification was considered satisfactory if the ratios of a quantifier and three most abundant characteristic ions reasonably matched those in the reference spectra of a given compound. Linear retention indices (relative to C7–C24 n-alkanes) were calculated and compared to those from literature. When standards were available, standard calibration curves based on quantifier ions were used for quantification. Linearity was satisfactory with coefficient of determination higher than 0.99 for all the standards.

For other compounds semi-quantitative analysis was carried out, and their concentrations (µg or mg/kg) were expressed as equivalents of the compounds with similar chemical structure for which standards were available, assuming a response factor equal to one.

2.5. Analysis of Phenols by UPLC-DAD

Extraction of phenols from EVOO was performed according to the modified method proposed by Jerman Klen, Golc Wondra, Vrhovšek, and Mozetič Vodopivec [42]. Ten grams of EVOO were dissolved in 10 mL of n-hexane, and 5 mL of methanol was added. The mixture was vortexed for 2 min, sonicated for 10 min, and then centrifuged at 5000 rpm for 5 min. The extraction was repeated 2 more times, and unified methanol extracts were defatted by 3 portions of 10 mL n-hexane. The methanol extracts (or standard solutions) were evaporated to dryness, the residue was re-dissolved in a 2 mL of a mixture of HPLC eluents (A (95:5 water—acetic acid (v/v)):B (methanol) = 90:10 (v/v)), and filtered through 0.45 µm PTFE filters.

Analysis of phenols was performed by ultra-performance liquid chromatography with diode array detection (UPLC-DAD) using an Agilent Infinity 1260 system (Agilent Technologies, Palo Alto, CA, USA) equipped with a G1311B quaternary pump, a G1329B autosampler, a G1316A column oven, and a G4212B DAD detector. A Kinetex PFP column (2.6 µm, 100 mm × 4.6 mm) with a guard was used (Phenomenex, Sydney, Australia) at 27 °C. Solvents were water with glacial acetic acid (95:5, v/v) (A) and methanol (B), with a flow rate of 1 mL/min. Ten microliters of the extract were injected. A 20-step gradient run used was reported previously [42]. Identification was performed by comparing retention times and UV/Vis spectra with those of pure standards when available, and with UV/Vis spectra from the literature [42]. Detection wavelengths were 280 nm (for simple phenols, vanillic acid, lignans, and secoiridoids), 320 nm (vanillin and *p*-coumaric acid), and 365 nm (flavonoids), while spectra were registered from 200 to 600 nm. Standard calibration curves were constructed for tyrosol, hydroxytyrosol, vanillic acid, vanillin, *p*-coumaric acid, luteolin, apigenin, pinoresinol, and oleuropein. For other compounds semi-quantitative analysis was carried out: secoiridoids were expressed in mg/kg as oleuropein, and acetoxypinoresinol as pinoresinol equivalents, respectively.

2.6. Statistical Data Elaboration

Data from GC-IT-MS, UPLC-DAD, and sensory analysis were subjected to one-way analysis of variance (ANOVA), and average values were compared by Least Significant Difference (LSD) test at the level of $p < 0.05$. Data were further processed by multivariate techniques, such as forward stepwise linear discriminant analysis (SLDA) and partial least squares discriminant analysis (PLSDA). The main goal of SLDA was to find the most useful variables (volatile compounds) for the mutual differentiation of all the six monovarietal EVOO. SLDA was applied on mean-centered data of a reduced dataset including six groups (varieties) and 50 variables with the highest F-ratios obtained in one-way ANOVA. Wilk's lambda was used as a selection criterion with an F statistic factor to establish the significance of the changes in Lambda when a new variable is tested (F-value to enter = 1). The main goal of PLSDA was to find the most useful variables (volatile compounds) for the differentiation of each of the six investigated monovarietal EVOO from all the other (five) monovarietal EVOOs. For this reason, PLSDA was applied on mean-centered data of six separate datasets each including two groups (a single vs. other five monovarietal EVOOs) and all the 197 variables. Variable Importance in Projection (VIP) scores were determined as the weighted sums of the squares of the weight in the PLSDA. ANOVA and SLDA data elaboration were performed by Statistica v. 13.2 software (StatSoft Inc., Tulsa, OK, USA), while PLSDA analysis was conducted using MetaboAnalyst v. 4.0 (http://www.metaboanalyst.ca) created at the University of Alberta, Canada [43].

3. Results

3.1. Volatile Aroma Compound Profiles

A total of 197 volatile compounds were reported, including 29 hydrocarbons, 29 terpenes, 24 aldehydes, 11 ketones, 23 alcohols, 10 acids, 17 esters, 37 benzenoids, 8 furanoids, and 9 other compounds (Table 1). For many volatiles significant differences between average concentrations in the investigated EVOOs were found. Several volatile compounds emerged as exclusive markers of particular monovarietal EVOOs.

3.1.1. Hydrocarbons

Istrian Buža and Rosinjola EVOOs stood out with the highest concentration of particular unsaturated hydrocarbons, such as several non-identified branched-chain alkenes, 3-ethyl-1,5-octadiene and 3,7-decadiene isomers, as well as that of saturated ones, such as decane, undecane, and dodecane (Table 1). On the other hand, lower amounts of the same groups of volatiles were found characteristic for Dalmatian Oblica and Lastovka, while I. bjelica EVOO contained intermediate concentrations. Similar relations were observed when comparing total hydrocarbons. Among hydrocarbons, 2,6-dimethyl-3-heptene turned out to be an exclusive marker of Rosinjola, and dodecene of Lastovka EVOO, respectively.

3.1.2. Monoterpenes and Sesquiterpenes

Lastovka EVOO was distinguished by the highest concentrations of several monoterpenes, such as α-pinene, camphene, myrcene, β-phellandrene, and γ-terpinene, as well as total monoterpenes (Table 1). The same EVOO contained the highest levels of γ-elemene and particular non-identified sesquiterpenes. Several sesquiterpenes were characteristic for Oblica EVOO, with α-muurolene as the most prominent marker. The lowest concentrations of (+)-cycloisosativene, α-muurolene, δ-cadinene, and two unidentified sesquiterpenes, as well as total sesquiterpenes, were found in Lastovka and Leccino EVOOs.

3.1.3. Aldehydes

Among unsaturated aldehydes formed in the so-called LOX pathway, Buža, followed by Rosinjola and Oblica EVOOs, contained the highest concentrations of (*E*)- and (*Z*)-3-hexenal, respectively (Table 1). The same monovarietal EVOOs, with sporadic exceptions, were also distinguished by high levels of pentenals, hexadienals, and (*E*)-2-octenal. Oblica EVOO had high concentration of decadienals. Leccino was clearly distinguished from the other monovarietal EVOOs by the highest level of the major EVOO volatile, (*E*)-2-hexenal, as well as that of (*E*,*E*)-2,4-heptadienal. Leccino EVOO had the lowest concentration of (*Z*)-3-hexenal, although not statistically different from that found in I. bjelica and Lastovka EVOOs. Lastovka EVOO was characterized by rather low levels of particular LOX-derived hexenals, as well as pentenals, octanal, and hexadienals. Istarska bjelica contained low concentrations of (*Z*)-3-hexenal and decadienals. Among saturated aldehydes originating from the processes other than LOX, 3-methylbutanal turned out to be an exclusive marker of I. bjelica, while abundance in 2-methyl-2-pentenal was observed in Rosinjola EVOO (Table 1). Higher levels of hexanal clearly discriminated Dalmatian (Oblica and Lastovka) from Istrian monovarietal EVOOs.

Table 1. Concentrations (µg/kg unless otherwise stated) of volatile aroma compounds determined by gas chromatography-ion trap mass spectrometry (GC-IT-MS) after headspace solid-phase microextraction (HS-SPME) from monovarietal extra virgin olive oils produced from Buža, Istarska bjelica, Rosinjola, Oblica, Lastovka, and Leccino varieties in Croatia.

Compound	ID	LRI	Variety					
			Buža	I. bjelica	Rosinjola	Oblica	Lastovka	Leccino
hydrocarbons								
propane	MS	<700	6.64 [bc]	14.91 [a]	1.12 [c]	11.49 [ab]	12.51 [a]	5.43 [c]
(Z)-2-pentene	MS	<700	61.19 [b]	80.27 [a]	60.31 [b]	33.98 [c]	43.63 [c]	83.09 [a]
1,4-pentadiene (mg/kg)	LRI,MS	<700	0.17 [a]	0.16 [a]	0.15 [ab]	0.10 [c]	0.09 [c]	0.12 [b]
1,3-pentadiene	LRI,MS	<700	81.73 [a]	75.36 [ab]	71.98 [ab]	53.80 [c]	48.98 [c]	62.92 [bc]
1-heptene	LRI,MS	731	5.42	16.64	3.25	9.76	5.72	6.13
3,5-dimethylheptane	MS	753	1.61	3.59	3.63	2.38	0.33	2.86
propanal	LRI,MS	778	7.00	5.23	7.36	5.10	5.68	7.04
octane	S,LRI,MS	800	31.03	35.33	27.29	30.78	40.92	27.11
2,6-dimethyl-3-heptene	MS	821	0.20 [b]	0.58 [b]	6.17 [a]	0.00 [b]	0.00 [b]	0.19 [b]
(E)-2-octene	LRI,MS	832	3.50	4.60	3.81	3.36	5.62	3.60
branched-chain alkene I (n.i.)	MS	953	94.32 [a]	63.73 [c]	85.98 [ab]	47.29 [d]	38.96 [d]	74.97 [bc]
branched-chain alkene II (n.i.)	MS	960	84.13 [a]	56.70 [c]	71.63 [b]	41.67 [d]	35.22 [d]	65.24 [bc]
decane	S,LRI,MS	997	31.53 [ab]	19.98 [bc]	39.92 [a]	9.49 [c]	7.39 [c]	14.67 [c]
3-ethyl-1,5-octadiene I (mg/kg)	LRI,MS	1006	0.36 [a]	0.26 [c]	0.34 [ab]	0.18 [d]	0.15 [d]	0.31 [b]
3-ethyl-1,5-octadiene II (mg/kg)	LRI,MS	1020	0.40 [a]	0.25 [c]	0.35 [ab]	0.20 [d]	0.16 [d]	0.31 [b]
3-ethyl-1,5-octadiene III	LRI,MS	1033	2.99	12.91	3.19	1.24	0.79	2.60
branched-chain alkene III (n.i.)	MS	1067	0.37	10.08	0.45	0.06	0.01	0.48
3,7-decadiene I	LRI,MS	1074	76.81 [a]	44.98 [c]	71.69 [ab]	33.06 [d]	29.07 [d]	65.14 [b]
branched-chain alkene IV (n.i.)	MS	1077	3.59	6.51	1.87	1.62	0.72	2.92
3,7-decadiene II (mg/kg)	LRI,MS	1080	0.29 [a]	0.17 [b]	0.28 [a]	0.15 [bc]	0.13 [c]	0.26 [a]
3,7-decadiene III (mg/kg)	LRI,MS	1084	0.29 [a]	0.15 [c]	0.25 [ab]	0.15 [c]	0.11 [c]	0.21 [b]
branched-chain alkene V (n.i.)	MS	1087	7.98 [ab]	11.87 [a]	6.30 [ab]	4.05 [b]	3.00 [b]	5.80 [b]
undecane	S,LRI,MS	1095	18.56 [ab]	9.75 [b]	39.12 [a]	3.29 [b]	2.09 [b]	9.83 [b]
dodecane	S,LRI,MS	1203	23.4 [ab]	14.16 [bc]	35.51 [a]	4.64 [c]	2.64 [c]	13.66 [bc]
dodecene	LRI,MS	1241	27.26 [bc]	33.95 [b]	17.25 [cd]	14.74 [cd]	58.52 [a]	8.43 [d]
3-propylcyclohexene	MS	1247	10.97 [bc]	7.93 [c]	14.15 [ab]	13.41 [ab]	9.90 [bc]	14.85 [a]
1,5,5,6-tetramethyl-1,3-cyclohexadiene	LRI,MS	1365	2.46 [bc]	1.55 [c]	4.38 [abc]	5.73 [a]	5.52 [ab]	1.82 [c]
2,2-dimethyl-(Z)-3-hexene I	MS	1499	28.41 [a]	16.09 [b]	33.60 [a]	26.67 [a]	14.60 [b]	10.76 [b]
2,2-dimethyl-(Z)-3-hexene II	MS	1560	4.79	2.12	6.05	4.78	2.39	2.88

Table 1. *Cont.*

Compound	ID	LRI	Variety					
			Buža	I. bjelica	Rosinjola	Oblica	Lastovka	Leccino
monoterpenes								
α-pinene	S,LRI,MS	1015	6.28 [b]	4.91 [b]	12.25 [b]	8.90 [b]	41.71 [a]	2.79 [b]
camphene	S,LRI,MS	1053	0.23 [b]	0.10 [b]	0.25 [b]	0.31 [b]	1.35 [a]	0.04 [b]
β-pinene	S,LRI,MS	1099	0.41	0.77	1.98	1.10	83.74	0.46
sabinene	S,LRI,MS	1112	2.07	0.79	2.09	2.94	49.44	17.20
3-carene	S,LRI,MS	1139	4.72 [a]	3.06 [b]	4.20 [ab]	0.97 [c]	2.29 [bc]	0.83 [c]
monoterpene I (n.i.)	MS	1140	2.77 [a]	1.84 [b]	2.54 [ab]	0.40 [c]	1.35 [bc]	0.27 [c]
myrcene	S,LRI,MS	1157	8.60 [b]	4.36 [b]	19.76 [ab]	11.48 [b]	41.67 [a]	11.85 [b]
α-terpinene	S,LRI,MS	1171	0.37	0.23	1.12	0.72	10.84	9.91
limonene (mg/kg)	S,LRI,MS	1191	0.02	0.08	0.02	0.02	0.64	0.01
β-phellandrene	LRI,MS	1201	0.00 [b]	0.00 [b]	0.00 [b]	0.03 [b]	4.25 [a]	0.02 [b]
(Z)-ocimene	S,LRI,MS	1230	15.14 [b]	8.00 [b]	25.27 [ab]	34.84 [a]	37.26 [a]	10.72 [b]
γ-terpinene	S,LRI,MS	1238	30.50 [b]	4.96 [b]	4.17 [b]	3.74 [b]	436.30 [a]	40.80 [b]
(E)-ocimene (mg/kg)	S,LRI,MS	1245	0.21 [bc]	0.09 [c]	0.27 [abc]	0.52 [a]	0.43 [ab]	0.12 [c]
monoterpene II (n.i.)	MS	1260	0.69 [bc]	0.40 [c]	1.28 [abc]	1.82 [a]	1.49 [ab]	0.53 [bc]
terpinolene	S,LRI,MS	1275	0.44	0.28	0.70	0.38	3.83	0.84
(Z)-alloocimene	LRI,MS	1349	11.38	9.68	10.20	23.02	8.22	7.20
(E)-alloocimene	S,LRI,MS	1357	16.79	12.63	13.25	37.22	10.66	9.26
linalool	S,LRI,MS	1536	15.44 [bc]	8.94 [c]	19.00 [abc]	31.43 [ab]	40.98 [a]	22.04 [abc]
sesquiterpenes								
(+)-cycloisosativene (mg/kg)	LRI,MS	1477	1.15 [a]	0.87 [b]	0.80 [b]	0.89 [ab]	0.24 [c]	0.26 [c]
α-copaene (mg/kg)	LRI,MS	1487	8.44 [a]	7.02 [ab]	5.81 [b]	8.77 [a]	2.14 [c]	1.64 [c]
sesquiterpene I (n.i.)	MS	1536	0.00 [b]	0.00 [b]	0.00 [b]	0.00 [b]	3.44 [a]	0.00 [b]
sesquiterpene II (n.i.) (mg/kg)	MS	1583	0.29 [a]	0.27 [a]	0.23 [a]	0.28 [a]	0.08 [b]	0.05 [b]
sesquiterpene III (n.i.) (mg/kg)	MS	1683	0.11 [ab]	0.09 [b]	0.09 [b]	0.13 [a]	0.03 [c]	0.02 [c]
δ-selinene	LRI,MS	1698	24.52 [b]	25.84 [ab]	10.32 [c]	32.29 [a]	6.82 [c]	2.07 [c]
γ-elemene (mg/kg)	MS	1704	0.04 [b]	0.05 [b]	0.05 [b]	0.06 [b]	0.20 [a]	0.05 [b]
α-muurolene (mg/kg)	LRI,MS	1719	1.65 [b]	1.42 [b]	1.26 [b]	2.10 [a]	0.49 [c]	0.34 [c]
sesquiterpene IV (n.i.)	MS	1736	0.63 [c]	3.02 [c]	26.21 [b]	2.33 [c]	51.54 [a]	29.75 [b]
α-farnesene (mg/kg)	LRI,MS	1745	0.53 [a]	0.29 [b]	0.84 [a]	0.12 [b]	0.25 [b]	0.17 [b]
δ-cadinene (mg/kg)	LRI,MS	1750	0.14 [ab]	0.11 [b]	0.17 [a]	0.17 [a]	0.06 [c]	0.07 [c]
aldehydes								
acrolein	LRI,MS	829	6.07 [a]	2.34 [d]	4.66 [ab]	4.23 [bc]	2.60 [cd]	3.75 [bc]

Table 1. *Cont.*

Compound	ID	LRI	Variety					
			Buža	I. bjelica	Rosinjola	Oblica	Lastovka	Leccino
2-methylbutanal	LRI,MS	903	10.74	13.39	11.93	7.86	18.70	13.76
3-methylbutanal	S,LRI,MS	906	10.51 [b]	22.29 [a]	10.65 [b]	8.49 [b]	12.31 [b]	12.26 [b]
2-methyl-2-pentenal	MS	930	1.08 [c]	0.26 [cd]	3.34 [a]	2.15 [b]	0.85 [cd]	0.00 [d]
(E)-2-butenal	S,LRI,MS	1024	11.44	9.98	12.87	9.10	18.39	25.45
unsaturated aliphatic aldehyde I (n.i.)	MS	1036	3.27 [a]	2.25 [b]	2.63 [ab]	2.48 [ab]	1.56 [b]	1.68 [b]
unsaturated aliphatic aldehyde II (n.i.)	MS	1051	0.25	5.31	0.55	0.03	0.04	0.25
hexanal (mg/kg)	S,LRI,MS	1070	0.24 [b]	0.29 [b]	0.27 [b]	0.44 [a]	0.53 [a]	0.20 [b]
(Z)-2-pentenal	LRI,MS	1093	11.73 [ab]	7.43 [c]	16.92 [a]	8.09 [bc]	5.29 [c]	5.70 [c]
(E)-2-pentenal	S,LRI,MS	1115	52.30 [a]	48.87 [a]	43.67 [ab]	39.40 [ab]	24.74 [b]	42.38 [ab]
(E)-3-hexenal	LRI,MS	1125	57.10 [a]	42.81 [b]	67.22 [a]	43.82 [ab]	23.48 [c]	35.30 [bc]
(Z)-3-hexenal (mg/kg)	LRI,MS	1130	1.36 [a]	0.40 [cd]	0.96 [abc]	1.09 [ab]	0.47 [bcd]	0.14 [d]
heptanal	LRI,MS	1175	4.37	3.60	4.33	3.19	3.03	3.90
(Z)-2-hexenal	LRI,MS	1189	44.49	31.63	48.50	33.38	39.16	32.68
(E)-2-hexenal (mg/kg)	S,LRI,MS	1205	19.38 [bc]	22.54 [b]	21.96 [b]	11.90 [cd]	5.92 [d]	34.95 [a]
octanal (mg/kg)	S,LRI,MS	1280	0.10 [a]	0.11 [a]	0.09 [a]	0.07 [b]	0.07 [b]	0.09 [a]
(Z)-2-heptenal	LRI,MS	1312	10.60	7.65	4.80	11.30	7.66	7.12
(E,E)-2,4-hexadienal (mg/kg)	LRI,MS	1381	0.27 [ab]	0.16 [cd]	0.35 [a]	0.20 [bc]	0.07 [d]	0.13 [cd]
(E,Z)-2,4-hexadienal (mg/kg)	LRI,MS	1385	1.56 [ab]	1.06 [bc]	2.19 [a]	0.87 [cd]	0.38 [d]	0.62 [cd]
(E)-2-octenal	S,LRI,MS	1417	11.75 [a]	5.53 [c]	8.12 [bc]	10.47 [ab]	7.17 [bc]	9.15 [b]
(E,E)-2,4-heptadienal	LRI,MS	1448	33.23 [b]	19.61 [d]	30.68 [bcd]	32.10 [bc]	21.83 [c]	44.72 [a]
2-isopropylidene-3-methylhexa-3,5-dienal	MS	1460	0.36 [bc]	0.33 [c]	0.65 [ab]	0.11 [c]	0.09 [c]	0.70 [a]
(E,E)-2,4-decadienal	LRI,MS	1750	0.65 [bc]	0.49 [c]	1.06 [ab]	1.52 [a]	0.85 [b]	0.85 [b]
(E,Z)-2,4-decadienal	LRI,MS	1794	0.67 [b]	0.26 [c]	0.47 [bc]	1.13 [a]	0.93 [ab]	0.61 [bc]
tetradecanal	LRI,MS	1909	1.05	0.97	1.04	1.41	0.99	0.62
ketones								
3-pentanone (mg/kg)	LRI,MS	962	0.09 [c]	0.23 [a]	0.09 [c]	0.13 [bc]	0.19 [ab]	0.08 [c]
1-penten-3-one (mg/kg)	S,LRI,MS	1008	0.29 [a]	0.29 [a]	0.26 [ab]	0.18 [bc]	0.12 [c]	0.22 [ab]
3-hexen-2-one	MS	1121	0.57 [b]	0.67 [a]	0.58 [b]	0.44 [c]	0.55 [b]	0.45 [c]
2-octanone	LRI,MS	1275	6.12	0.62	0.54	1.55	1.51	0.89
2-methyl-6-methylene-1,7-octadien-3-one	LRI,MS	1303	44.85 [a]	21.50 [b]	51.45 [a]	24.25 [b]	16.32 [b]	62.59 [a]
6-methyl-5-hepten-2-one	LRI,MS	1326	2.15 [bc]	1.94 [bc]	2.63 [abc]	3.02 [ab]	4.40 [a]	1.40 [c]
2-cyclohexene-1,4-dione I	MS	1712	3.62 [ab]	2.27 [cd]	5.11 [a]	3.20 [bc]	2.05 [bcd]	1.04 [d]
2-cyclohexene-1,4-dione II	MS	1803	1.15 [ab]	0.68 [cd]	1.57 [a]	1.06 [abc]	0.64 [cd]	0.29 [d]

Table 1. *Cont.*

Compound	ID	LRI	Variety					
			Buža	I. bjelica	Rosinjola	Oblica	Lastovka	Leccino
(Z)-cinerolone	MS	2002	0.20 [b]	0.13 [c]	0.19 [b]	0.04 [d]	0.03 [d]	0.28 [a]
1-(2,6,6-trimethyl-1-cyclohexen-1-yl)-1-penten-3-one	MS	2056	0.18 [c]	0.25 [bc]	0.14 [c]	0.51 [a]	0.47 [ab]	0.12 [c]
4′-ethoxy-2′-hydroxyoctadecanophenone	MS	2081	0.26 [bc]	0.12 [c]	0.90 [a]	0.24 [bc]	0.26 [bc]	0.41 [b]
alcohols								
2-methyl-2-propanol	LRI,MS	886	4.98	4.93	8.45	0.03	0.00	4.33
1-methoxy-2-propanol (mg/kg)	MS	917	0.15 [b]	0.12 [b]	0.18 [ab]	0.15 [b]	0.23 [a]	0.14 [b]
3-pentanol	LRI,MS	1097	11.84	16.51	6.53	6.20	15.09	5.92
2-pentanol	LRI,MS	1109	1.04	2.75	0.67	2.76	1.78	1.59
1-penten-3-ol (mg/kg)	LRI,MS	1148	0.18 [b]	0.28 [a]	0.19 [b]	0.14 [b]	0.19 [b]	0.18 [b]
3-methyl-1-butanol	S,LRI,MS	1197	19.96 [b]	26.70 [b]	22.74 [b]	27.21 [b]	57.71 [a]	23.80 [b]
1-pentanol	S,LRI,MS	1239	10.37 [bc]	16.24 [ab]	7.86 [bc]	20.10 [a]	22.07 [a]	6.96 [c]
(E)-2-penten-1-ol	S,LRI,MS	1300	62.07	65.28	64.27	43.81	67.92	61.24
(Z)-2-penten-1-ol (mg/kg)	S,LRI,MS	1308	0.26 [a]	0.31 [a]	0.27 [a]	0.21 [b]	0.28 [a]	0.25 [ab]
1-hexanol (mg/kg)	S,LRI,MS	1342	0.70	1.21	1.20	1.17	1.58	0.76
(E)-3-hexen-1-ol	LRI,MS	1352	27.42 [ab]	27.45 [ab]	27.49 [ab]	37.90 [a]	46.03 [a]	14.76 [b]
(Z)-3-hexen-1-ol (mg/kg)	S,LRI,MS	1372	1.63 [b]	1.08 [bc]	1.50 [bc]	2.77 [a]	1.92 [ab]	0.61 [c]
(E)-2-hexen-1-ol (mg/kg)	S,LRI,MS	1394	1.00 [b]	1.68 [b]	1.37 [b]	1.05 [b]	3.42 [a]	1.42 [b]
(Z)-2-hexen-1-ol	S,LRI,MS	1402	7.53	122.27	11.84	11.20	15.53	10.63
1-heptanol	S,LRI,MS	1445	4.05	2.16	2.23	3.38	3.39	2.71
2-ethyl-1-hexanol	S,LRI,MS	1480	4.58 [d]	5.90 [c]	10.39 [a]	4.32 [d]	4.01 [d]	6.93 [b]
1-octanol	S,LRI,MS	1546	7.65	5.29	5.41	5.56	6.01	5.98
2,4-hexadien-1-ol	LRI,MS	1568	0.11	0.10	0.00	0.65	0.64	0.00
1-nonanol	S,LRI,MS	1650	3.73	2.56	3.14	2.53	3.32	3.10
3-methyl-3-cyclohexen-1-ol	MS	1744	4.66 [a]	2.34 [b]	4.39 [a]	2.25 [b]	0.65 [b]	0.88 [b]
2-(2-butoxyethoxy)-ethanol	LRI,MS	1779	3.83 [b]	2.43 [b]	42.91 [a]	7.86 [b]	7.62 [b]	9.91 [b]
2,2′-oxybis-1-propanol	LRI,MS	1865	4.04	3.46	4.51	2.27	2.59	2.44
tetradecanol	LRI,MS	2158	25.23 [a]	26.72 [a]	12.11 [ab]	31.35 [a]	13.64 [ab]	6.70 [b]
acids								
acetic acid (mg/kg)	S,LRI,MS	1427	1.82	2.55	2.46	2.49	4.53	1.70
acid (n.i.) (mg/kg)	MS	1476	0.91 [ab]	0.82 [b]	1.17 [a]	1.10 [a]	1.18 [a]	1.07 [a]
butanoic acid	S,LRI,MS	1602	27.65 [c]	41.57 [bc]	40.46 [bc]	52.13 [ab]	72.47 [a]	30.43 [bc]
hexanoic acid (mg/kg)	S,LRI,MS	1823	0.76 [c]	1.23 [bc]	1.32 [bc]	1.74 [ab]	2.21 [a]	1.00 [c]
2-ethylhexanoic acid	LRI,MS	1925	3.41 [c]	8.59 [c]	52.77 [a]	10.47 [bc]	12.29 [bc]	23.46 [b]
(E)-3-hexenoic acid	LRI,MS	1926	12.93 [b]	9.33 [bc]	16.87 [ab]	24.46 [a]	5.08 [bc]	1.38 [c]

Foods **2019**, *8*, 565

Table 1. *Cont.*

Compound	ID	LRI	Variety					
			Buža	I. bjelica	Rosinjola	Oblica	Lastovka	Leccino
octanoic acid (mg/kg)	S,LRI,MS	2033	0.48 [b]	7.82 [a]	1.29 [b]	1.65 [b]	1.47 [b]	0.59 [b]
sorbic acid	LRI,MS	2048	7.83 [b]	4.94 [bc]	14.78 [a]	8.44 [b]	4.58 [bc]	1.04 [c]
nonanoic acid (mg/kg)	S,LRI,MS	2139	41.76 [b]	434.34 [a]	54.30 [b]	41.10 [b]	55.83 [b]	29.51 [b]
decanoic acid (mg/kg)	S,LRI,MS	2244	0.07 [b]	0.43 [a]	0.14 [b]	0.02 [b]	0.06 [b]	0.06 [b]
esters								
allyl acetate	LRI,MS	805	53.61	58.55	69.96	43.21	30.69	37.81
methyl acetate	LRI,MS	816	91.00 [b]	166.56 [a]	101.38 [ab]	59.88 [b]	115.21 [ab]	81.99 [b]
1,1,1-trimethoxyethane	MS	871	61.25 [b]	328.53 [a]	117.33 [ab]	50.70 [b]	85.45 [b]	24.70 [b]
ethyl 2-methylbutanoate	LRI,MS	1042	0.90	0.64	0.55	0.70	0.70	0.51
isoamyl acetate	S,LRI,MS	1114	4.91 [a]	4.92 [a]	2.26 [b]	2.88 [b]	3.19 [b]	1.69 [b]
methyl 3-methyl-2-butenoate	LRI,MS	1157	3.79	0.85	2.84	2.61	2.86	1.75
methyl hexanoate	LRI,MS	1178	1.29	1.01	2.05	1.51	1.02	1.29
ethyl hexanoate	S,LRI,MS	1228	0.06	0.23	0.31	0.50	0.35	0.12
hexyl acetate	S,LRI,MS	1265	21.75 [b]	89.41 [a]	25.62 [b]	11.18 [b]	91.10 [a]	13.11 [b]
(Z)-3-hexen-1-yl acetate (mg/kg)	S,LRI,MS	1309	0.27 [b]	0.62 [a]	0.20 [b]	0.05 [b]	0.26 [b]	0.09 [b]
(Z)-2-hexen-1-yl acetate	LRI,MS	1326	1.52 [b]	4.02 [a]	1.44 [b]	0.91 [b]	1.75 [b]	0.91 [b]
(E)-3-hexenyl butanoate	LRI,MS	1454	2.23 [abc]	0.88 [c]	2.34 [ab]	1.47 [bc]	2.94 [a]	1.71 [abc]
3-hydroxy-2,4,4-trimethylpentyl 2-methylpropanoate	MS	1854	0.37 [c]	0.16 [c]	4.94 [a]	1.06 [bc]	0.95 [bc]	1.54 [b]
methyl cinnamoylglycinate	MS	1960	3.44 [b]	3.34 [b]	6.22 [a]	2.62 [c]	2.25 [c]	3.33 [b]
triacetin	LRI,MS	2049	1.09 [bc]	0.41 [c]	5.19 [a]	1.44 [bc]	1.02 [bc]	1.77 [b]
2-propenyl pentanoate	MS	2074	3.07 [bc]	1.38 [cd]	7.81 [a]	3.79 [b]	2.39 [bcd]	0.80 [d]
methyl 3-oxo-2-pentyl-cyclopentaneacetate	MS	2257	7.59 [ab]	3.69 [c]	12.29 [a]	8.44 [ab]	8.06 [ab]	6.03 [bc]
benzenoids								
benzene	LRI,MS	926	9.83	14.67	14.44	12.05	11.62	10.10
toluene (mg/kg)	LRI,MS	1027	0.10	0.17	0.07	0.40	0.17	0.10
m-xylene	LRI,MS	1122	22.55	19.69	15.30	105.52	26.14	17.65
p-xylene (mg/kg)	LRI,MS	1128	0.06	0.06	0.05	0.22	0.07	0.05
o-xylene (mg/kg)	LRI,MS	1172	0.03	0.03	0.02	0.11	0.04	0.03
p-ethyltoluene	LRI,MS	1213	16.73	17.37	11.75	56.70	14.44	11.29
1,3,5-trimethylbenzene (mesitylene)	LRI,MS	1235	11.77	11.52	12.45	32.07	9.69	8.84
2-ethyltoluene	LRI,MS	1251	7.00	6.28	5.39	18.11	5.10	4.07
p-cymene	S,LRI,MS	1262	1.51	1.57	2.15	2.87	31.68	4.46
m-cymene	LRI,MS	1270	19.46	16.99	16.47	40.75	10.97	10.45
1,3-diethylbenzene	LRI,MS	1293	1.54	0.75	0.84	2.14	0.59	0.71

Table 1. *Cont.*

Compound	ID	LRI	Variety					
			Buža	I. bjelica	Rosinjola	Oblica	Lastovka	Leccino
o-cymene	LRI,MS	1296	1.64	1.38	1.37	3.62	0.94	1.09
1-ethyl-3,5-dimethylbenzene	LRI,MS	1316	4.25	4.17	3.80	8.75	2.79	3.11
1,2,3-trimethylbenzene (hemellitol)	LRI,MS	1324	8.25	6.56	7.35	14.88	4.70	4.04
anisole	LRI,MS	1329	3.77	2.45	2.91	1.27	3.42	1.85
1,2,4,5-tetramethylbenzene	LRI,MS	1413	1.89	1.36	1.44	3.99	1.13	1.09
isodurene	LRI,MS	1422	2.78	2.21	2.22	5.44	2.06	1.80
1,2,3,4-tetramethylbenzene	MS	1423	2.78	2.20	2.20	5.39	2.01	1.81
p-cymenene	LRI,MS	1436	9.05 [b]	7.21 [c]	13.45 [a]	7.49 [c]	6.48 [c]	6.87 [c]
methyl benzoate	LRI,MS	1601	14.74 [b]	3.88 [c]	22.08 [b]	41.94 [a]	13.97 [b]	4.01 [c]
acetophenone	S,LRI,MS	1626	14.76 [b]	13.46 [b]	20.73 [a]	13.08 [b]	12.68 [b]	13.60 [b]
estragole	LRI,MS	1655	0.06 [c]	0.03 [c]	0.02 [c]	0.90 [b]	1.40 [a]	0.01 [c]
4-ethylbenzaldehyde	LRI,MS	1689	6.13 [b]	4.73 [c]	9.68 [a]	4.66 [c]	3.77 [c]	4.66 [c]
methyl salicylate	LRI,MS	1755	47.45 [ab]	18.17 [d]	62.13 [a]	61.68 [a]	37.24 [bc]	30.00 [c]
aromatic aldehyde I (n.i.)	MS	1810	9.56 [b]	9.57 [b]	18.54 [a]	6.75 [c]	5.56 [c]	9.18 [c]
aromatic aldehyde II (n.i.)	MS	1839	9.12 [b]	9.29 [b]	18.39 [a]	6.46 [c]	5.50 [c]	8.81 [c]
benzyl alcohol	S,LRI,MS	1851	20.21 [c]	22.03 [c]	24.09 [bc]	23.65 [bc]	35.74 [ab]	39.17 [a]
2-phenylethyl alcohol (mg/kg)	S,LRI,MS	1885	0.33	0.30	0.39	0.35	0.32	0.38
benzyl nitrile	LRI,MS	1896	0.04 [b]	0.00 [b]	0.16 [b]	0.00 [b]	0.00 [b]	1.17 [a]
lilial	LRI,MS	2020	1.55 [bc]	1.58 [bc]	1.04 [bc]	2.42 [ab]	3.74 [a]	1.01 [c]
methyl 2-methoxybenzoate	LRI,MS	2037	5.21 [a]	0.43 [c]	4.21 [ab]	3.49 [ab]	2.45 [bc]	1.19 [c]
complex benzenoid (n.i.)	MS	2041	3.94 [b]	5.40 [b]	3.69 [b]	12.24 [a]	11.25 [a]	3.07 [b]
2-phenoxyethanol	LRI,MS	2106	3.30 [bc]	2.07 [c]	4.99 [bc]	10.29 [a]	7.11 [ab]	5.49 [bc]
methyl anthranilate	LRI,MS	2200	0.73 [b]	0.05 [c]	0.76 [b]	1.15 [b]	2.04 [a]	0.01 [c]
menthyl salicylate	MS	2273	1.25 [b]	2.57 [a]	2.16 [ab]	2.74 [a]	2.27 [a]	1.39 [b]
4-ethoxystyrene	MS	2348	1.97 [c]	1.96 [c]	2.20 [bc]	3.07 [a]	2.74 [ab]	1.95 [c]
benzoic acid	LRI,MS	2381	11.32 [b]	44.78 [a]	38.56 [a]	5.62 [b]	7.75 [b]	14.51 [b]
furanoids								
2-ethylfuran (mg/kg)	LRI,MS	938	0.14 [b]	0.08 [cd]	0.22 [a]	0.09 [bc]	0.05 [cd]	0.03 [d]
2-vinylfuran	LRI,MS	1059	10.26 [b]	5.89 [bc]	17.85 [a]	9.15 [b]	5.32 [bc]	1.94 [c]
4-methyl-2,3-dihydrofuran	LRI,MS	1184	4.98 [ab]	4.59 [b]	3.83 [b]	6.13 [a]	4.90 [ab]	3.86 [b]
2-pentylfuran	LRI,MS	1225	7.81	7.60	7.33	8.82	10.79	62.68
5-methyl-2-furancarboxaldehyde	LRI,MS	1551	4.99 [b]	3.33 [c]	7.19 [a]	3.58 [bc]	1.90 [cd]	1.55 [d]
2(5H)-furanone	LRI,MS	1722	5.90	3.67	5.23	2.23	1.42	2.94

Table 1. *Cont.*

Compound	ID	LRI	Variety					
			Buža	I. bjelica	Rosinjola	Oblica	Lastovka	Leccino
5-ethyl-2(5H)-furanone (mg/kg)	LRI,MS	1733	0.19 [a]	0.08 [bc]	0.19 [a]	0.15 [ab]	0.06 [bc]	0.03 [c]
2-ethyl-5-methyl-tetrahydrofuran (mg/kg)	MS	1933	0.44 [ab]	0.21 [cd]	0.58 [a]	0.34 [bc]	0.15 [cd]	0.07 [d]
miscellaneous								
dimethyl sulfide	LRI,MS	739	1.21 [ab]	1.26 [ab]	2.49 [a]	1.18 [ab]	0.74 [b]	1.81 [ab]
2,4-dihydro-5-methyl-3H-pyrazol-3-one I	MS	978	0.90	2.14	1.10	10.00	4.62	0.38
1,2-dihydro-5-methyl-3H-pyrazol-3-one	MS	1109	0.85	1.17	0.52	4.51	2.12	0.09
2,4-dihydro-5-methyl-3H-pyrazol-3-one II	MS	1150	0.08	0.35	0.09	1.82	0.60	0.01
2-phenyl-1H-indole	MS	1502	78.56 [c]	93.8 [b]	139.59 [a]	67.51 [c]	74.11 [c]	95.38 [b]
dimethyl sulfoxide	LRI,MS	1540	26.39 [abc]	15.52 [c]	27.62 [ab]	32.02 [a]	23.96 [abc]	16.65 [bc]
n.i. (*m/z* 189,207,131)	MS	1688	0.51 [a]	0.21 [b]	0.45 [ab]	0.47 [a]	0.25 [ab]	0.19 [b]
n.i. (*m/z* 84,85,41,42,39,133,147,175)	MS	1990	1.91 [b]	1.20 [b]	3.72 [a]	1.07 [b]	1.32 [b]	1.21 [b]
phenol	S,LRI,MS	1972	5.27 [c]	6.01 [b]	9.88 [a]	4.38 [d]	4.16 [d]	5.95 [b]
totals (mg/kg)								
total hydrocarbons			2.12 [a]	1.54 [b]	1.99 [a]	1.13 [c]	1.01 [c]	1.70 [b]
total monoterpenes			0.35 [b]	0.23 [b]	0.41 [b]	0.71 [b]	1.85 [a]	0.26 [b]
total sesquiterpenes			12.36 [a]	10.15 [a]	9.29 [a]	12.54 [a]	3.54 [b]	2.66 [b]
total aldehydes			23.19 [b]	24.78 [b]	26.10 [b]	14.79 [c]	7.64 [c]	36.37 [a]
total ketones			0.44 [b]	0.56 [a]	0.41 [bc]	0.34 [c]	0.33 [c]	0.37 [c]
total alcohols			4.12 [bc]	5.02 [bc]	4.95 [bc]	5.69 [ab]	7.89 [a]	3.52 [c]
total acids			45.85 [b]	447.25 [a]	60.81 [b]	48.19 [b]	65.37 [b]	33.99 [b]
total esters			0.52 [b]	1.29 [a]	0.56 [b]	0.25 [b]	0.61 [b]	0.27 [b]
total benzenoids			0.80	0.81	0.87	1.60	0.88	0.79
total furanoids			0.80	0.40	1.04	0.61	0.28	0.76
total miscellaneous			0.12 [b]	0.12 [b]	0.19 [a]	0.12 [b]	0.11 [b]	0.12 [b]

ID—identification of compounds; S—retention time and mass spectrum consistent with that of the pure standard and with NIST05 mass spectra electronic library; LRI—linear retention index consistent with that found in literature; MS—mass spectra consistent with that from NIST05 mass spectra electronic library or literature; n.i.—not identified. The compounds with only MS symbol in ID column were tentatively identified. The compounds for which pure standards were not available (without symbol S in the ID column) were quantified semi-quantitatively and their concentrations were expressed as equivalents of compounds with similar chemical structure assuming a response factor = 1. Different superscript lowercase letters in a row represent statistically significant differences between mean values at $p < 0.05$ obtained by one-way ANOVA and least significant difference (LSD) test.

3.1.4. Ketones

Istrian EVOOs contained higher concentration of the most important olive oil ketone in sensory terms, 1-penten-3-one, in relation to the Dalmatian ones, especially Lastovka (Table 1). The highest concentrations of 2-cyclohehene-1,4-dione were found in Rosinjola followed by Buža, while the lowest were found in Leccino EVOO. Dalmatian Oblica and Lastovka EVOOs were distinguished by the highest concentration of 1-(2,6,6-trimethyl-1-cyclohexene-1-yl)-1-penten-3-one and the lowest concentration of (Z)-cinerolone. The latter volatile compound was found to be a marker of Leccino, the same as 4′-ethoxy-2′-hydroxyoctadecanophenone was for Rosinjola EVOO.

3.1.5. Alcohols

Among LOX-generated unsaturated C6 alcohols, a similar pattern as in the case of 3-hexenals was observed, with Leccino EVOO containing the lowest concentration of both (E)- and (Z)-3-hexen-1-ol (Table 1). Dalmatian EVOOs, especially Lastovka in the case of (E)-2-hexen-1-ol, exhibited the highest concentrations. 1-Penten-3-ol turned out to be a marker of I. bjelica EVOO, Rosinjola EVOO was the most abundant in 2-ethyl-1-hexanol and 2-(2-butoxyethoxy)-ethanol, while higher concentration of a number of non-LOX alcohols, such as 1-methoxy-2-propanol, 3-methyl-1-butanol, and 1-pentanol, turned out to be a feature of Lastovka EVOO.

3.1.6. Acids

Leccino EVOO was characterized by the lowest concentration of (E)-3-hexenoic acid (Table 1). Dalmatian, especially Lastovka EVOO, had the highest concentrations of butanoic and hexanoic acids, while I. bjelica EVOO was by far the most abundant in other middle-chain volatile fatty acids, especially nonanoic acid, as well as total acids. Rosinjola was distinguished by higher levels of 2-ethylhexanoic acid, which corresponded well to the higher concentration of 2-ethyl-1-hexanol found in this EVOO.

3.1.7. Esters

Istarska bjelica EVOO exhibited the highest concentration of the acetates of C6 alcohols, methyl acetate, and total esters (Table 1). Lastovka EVOO was abundant in hexyl acetate. Rosinjola EVOO stood out with the highest levels of several, mostly tentatively identified esters with high LRIs.

3.1.8. Benzenoids

For many simple benzenoids (e.g., xylenes, ethylbenzenes, cymenes, etc.) no statistically significant differences between the monovarietal EVOOs were observed (Table 1). Rosinjola was the most distinguished by the highest concentrations of *p*-cymenene, acetophenone, 4-ethylbenzaldehyde, and the two non-identified aromatic aldehydes. Estragole and methyl anthranilate were found in the highest concentration in Lastovka EVOO, and these two compounds, together with lilial, 2-phenoxyethanol, and 4-ethoxystyrene, were more abundant in Dalmatian than in Istrian EVOOs. The highest concentration of benzyl nitrile was observed in Leccino EVOO. Low concentrations of several benzenoids were characteristic for particular EVOOs: methyl salicylate in I. bjelica, and methyl 2-methoxybenzoate and methyl anthranilate in I. bjelica and Leccino.

3.1.9. Furanoids

Similar as for the benzenoids, Rosinjola EVOO was characterized by several furanoid markers, including 2-ethylfuran, 2-vinylfuran, and 5-methyl-2-furancarboxaldehyde (Table 1). On the other hand, Leccino EVOO had the lowest concentration of furanoids in general.

3.1.10. Miscellaneous Compounds

Rosinjola EVOO contained the highest concentration of 2-phenyl-1H-indole and phenol (Table 1). Phenol concentration was the lowest in Dalmatian Oblica and Lastovka EVOOs.

3.1.11. Odor Activity Values (OAV)

Table 2 lists the average odor activity values (OAV) of the volatile aroma compounds found in the investigated EVOOs, calculated as the ratios of their concentrations and odor perception thresholds available in literature. For thirteen compounds average OAV higher than 1 was observed in at least one of the monovarietal EVOOs implying their direct influence on the aroma. The compound with the highest OAV was (Z)-3-hexenal, followed by 1-penten-3-one and (E)-2-hexenal, while other compounds exhibited much lower OAVs. (Z)-3-hexenal was potentially the most important odorant in all the investigated monovarietal EVOOs except I. bjelica and Leccino in which 1-penten-3-one was dominant.

Table 2. Sensory descriptors and odor perception thresholds of volatile aroma compounds in monovarietal extra virgin olive oils produced from Buža, Istarska bjelica, Rosinjola, Oblica, Lastovka, and Leccino varieties in Croatia, sorted in descending order according to their average odor activity values (OAV).

Volatile Compound	Sensory Descriptor (Aroma) *	Threshold *	Odor Activity Value (OAV)					
			Buža	I. bjelica	Rosinjola	Oblica	Lastovka	Leccino
OAV > 1								
(Z)-3-hexenal	leaf-like, green, apple-like	1.7	**800.00** [a]	235.29 [cd]	564.71 [abc]	641.18 [ab]	276.47 [bcd]	82.35 [d]
1-penten-3-one	leaf, green, pungent, sweet	0.73	**397.26** [a]	**397.26** [a]	356.16 [ab]	246.58 [bc]	164.38 [c]	301.37 [ab]
(E)-2-hexenal	green, apple-like, bitter almond	420	46.14 [bc]	53.67 [b]	52.29 [b]	28.33 [cd]	14.10 [d]	**83.21** [a]
hexanal	green, sweet, green apple, grassy	75	3.20 [b]	3.87 [b]	3.60 [b]	5.87 [a]	**7.07** [a]	2.67 [b]
3-methylbutanal	malty	5.2	2.02 [b]	**4.29** [a]	2.05 [b]	1.63 [b]	2.37 [b]	2.36 [b]
1-hexanol	fruit, banana, soft, grass	400	1.75	3.03	3.00	2.93	**3.95**	1.90
2-methylbutanal	malty	5.4	1.99	2.48	2.21	1.46	**3.46**	2.55
hexanoic acid	pungent, rancid, sweaty	700	1.09 [c]	1.76 [bc]	1.89 [bc]	2.49 [ab]	**3.16** [a]	1.43 [c]
(Z)-3-hexen-1-yl acetate	green, banana-like, olive fruity	200	1.35 [b]	**3.10** [a]	1.00 [b]	0.25 [b]	1.30 [b]	0.45 [b]
(E)-2-octenal	herbaceous, spicy	4	**2.94** [a]	1.38 [c]	2.03 [bc]	2.62 [ab]	1.79 [bc]	2.29 [b]
octanoic acid	oily, fatty	3000	0.16 [b]	**2.61** [a]	0.43 [b]	0.55 [b]	0.49 [b]	0.20 [b]
(Z)-3-hexen-1-ol	green, apple, leaf-like, banana	1100	1.48 [b]	0.98 [bc]	1.36 [bc]	**2.52** [a]	1.75 [ab]	0.55 [c]
ethyl 2-methylbutanoate	fruity	0.72	**1.25**	0.89	0.76	0.97	0.97	0.71
OAV < 1								
1-penten-3-ol	lawn, olive, leaf, pungent	400	0.45 [b]	**0.70** [a]	0.48 [b]	0.35 [b]	0.48 [b]	0.45 [b]
(E)-2-hexen-1-ol	green, grass, leaves, sweet	5000	0.20 [b]	0.34 [b]	0.27 [b]	0.21 [b]	**0.68** [a]	0.28 [b]
3-methyl-1-butanol	woody, whiskey, sweet	100	0.20 [b]	0.27 [b]	0.23 [b]	0.27 [b]	**0.58** [a]	0.24 [b]
octanal	fatty, sharp, citrus-like, soapy	320	0.31 [a]	**0.34** [a]	0.28 [a]	0.22 [b]	0.22 [b]	0.28 [a]
(E)-2-penten-1-ol	green fruity, fresh olive fruits	250	0.25	0.26	0.26	0.18	**0.27**	0.24
(E)-2-pentenal	green, apple, bitter almond	300	**0.17** [a]	0.16 [a]	0.15 [ab]	0.13 [ab]	0.08 [b]	0.14 [ab]
butanoic acid	rancid, cheese	650	0.04 [c]	0.06 [bc]	0.06 [bc]	0.08 [ab]	**0.11** [a]	0.05 [bc]
(E,Z)-2,4-decadienal	deep-fried	10	0.07 [b]	0.03 [c]	0.05 [bc]	**0.11** [a]	0.09 [ab]	0.06 [bc]
hexyl acetate	green, fruity, sweet, apple	1040	0.02 [b]	**0.09** [a]	0.02 [b]	0.01 [b]	**0.09** [a]	0.01 [b]
1-pentanol	fruity, strong, sticky, balsamic	470	0.02 [bc]	0.03 [ab]	0.02 [bc]	0.04 [a]	**0.05** [a]	0.01 [c]
octane	sweety, alcane	940	0.03	**0.04**	0.03	0.03	**0.04**	0.03
2-octanone	mould, green	510	**0.01**	0.00	0.00	0.00	0.00	0.00
(E)-3-hexen-1-ol	green, bitter	1500	0.02 [ab]	0.02 [ab]	0.02 [ab]	**0.03** [a]	**0.03** [a]	0.01 [b]
1-nonanol	fatty, rancid	280	0.01	0.01	0.01	0.01	0.01	0.01
heptanal	oily, fatty, woody	500	0.01	0.01	0.01	0.01	0.01	0.01
(E,E)-2,4-heptadienal	fatty, rancid	3620	0.01 [b]	0.01 [d]	0.01 [bcd]	0.01 [bc]	0.01 [c]	0.01 [a]
(E,E)-2,4-decadienal	deep-fried	180	0.00 [bc]	0.00 [c]	0.01 [ab]	0.01 [a]	0.00 [b]	0.00 [b]
6-methyl-5-hepten-2-one	pungent, green	1000	0.00 [bc]	0.00 [bc]	0.00 [abc]	0.00 [ab]	0.00 [a]	0.00 [c]
3-pentanone	fruity, green, sweet	70,000	0.00 [c]	0.00 [a]	0.00 [c]	0.00 [bc]	0.00 [ab]	0.00 [c]

* sensory descriptors and odor perception thresholds (µg/kg oil) reported in literature [21,44–47]. Values in bold indicate the highest average OAV for a given volatile compound among monovarietal extra virgin olive oils.

3.1.12. Multivariate Statistical Analysis

A differentiation model built by SLDA classified correctly all the monovarietal EVOOs according to variety (Figure 1) and extracted 30 variables (Table S2). A 100% correct classification was obtained after including 22 variables. Phenol was included in the model as the first and classified correctly 41.76% of all the investigated EVOO samples. After subsequently including nonanoic acid, α-muurolene, 3,7-decadiene I, and estragole (five compounds in total), the total percentage of the correctly classified EVOOs increased to 94.51%.

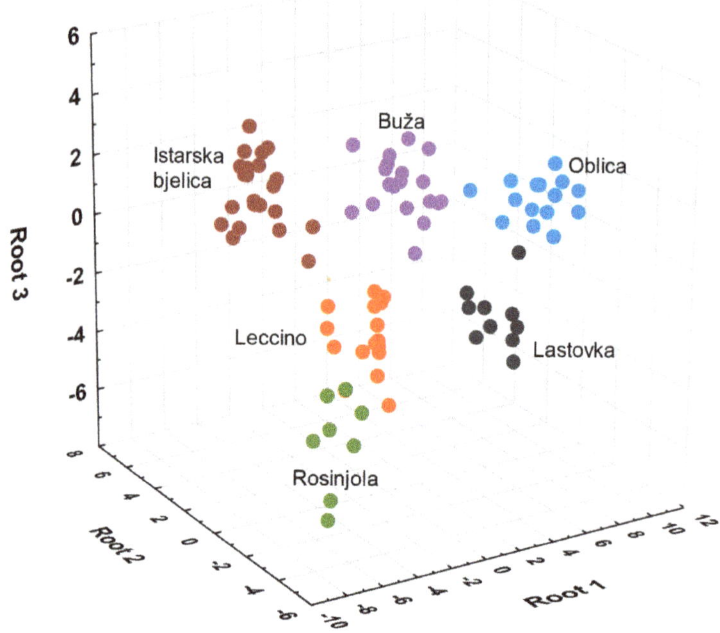

Figure 1. Separation of monovarietal extra virgin olive oils produced from Buža, Istarska bjelica, Rosinjola, Oblica, Lastovka, and Leccino varieties in Croatia according to variety in three-dimensional space defined by the first three discriminant functions (roots) on the basis of volatile aroma compound composition.

Unsaturated hydrocarbons from the LOX pathway, accompanied by the most important LOX volatile (Z)-3-hexenal, were characterized by the highest positive VIP scores obtained by PLSDA and were confirmed to be typical for Buža EVOO (Figure 2a). Middle-chain fatty acids and hexenol acetates were the volatiles with the highest VIP scores in I. bjelica EVOO (Figure 2b), while the markers of Rosinjola EVOO were mostly benzenoids (Figure 2c). The volatile compound with by far the highest VIP score for the discrimination of Oblica was methyl benzene (Figure 2d), while Lastovka EVOO typicity was mostly owed to its abundance in terpenes and deficiency in LOX volatiles (Figure 2e). High concentrations of benzyl nitrile, (E)-2-hexenal, and (E,E)-2,4-heptadienal, as well as low concentrations of sesquiterpenes, were confirmed to be the most prominent typical characteristics of the Leccino EVOO volatile profile (Figure 2f).

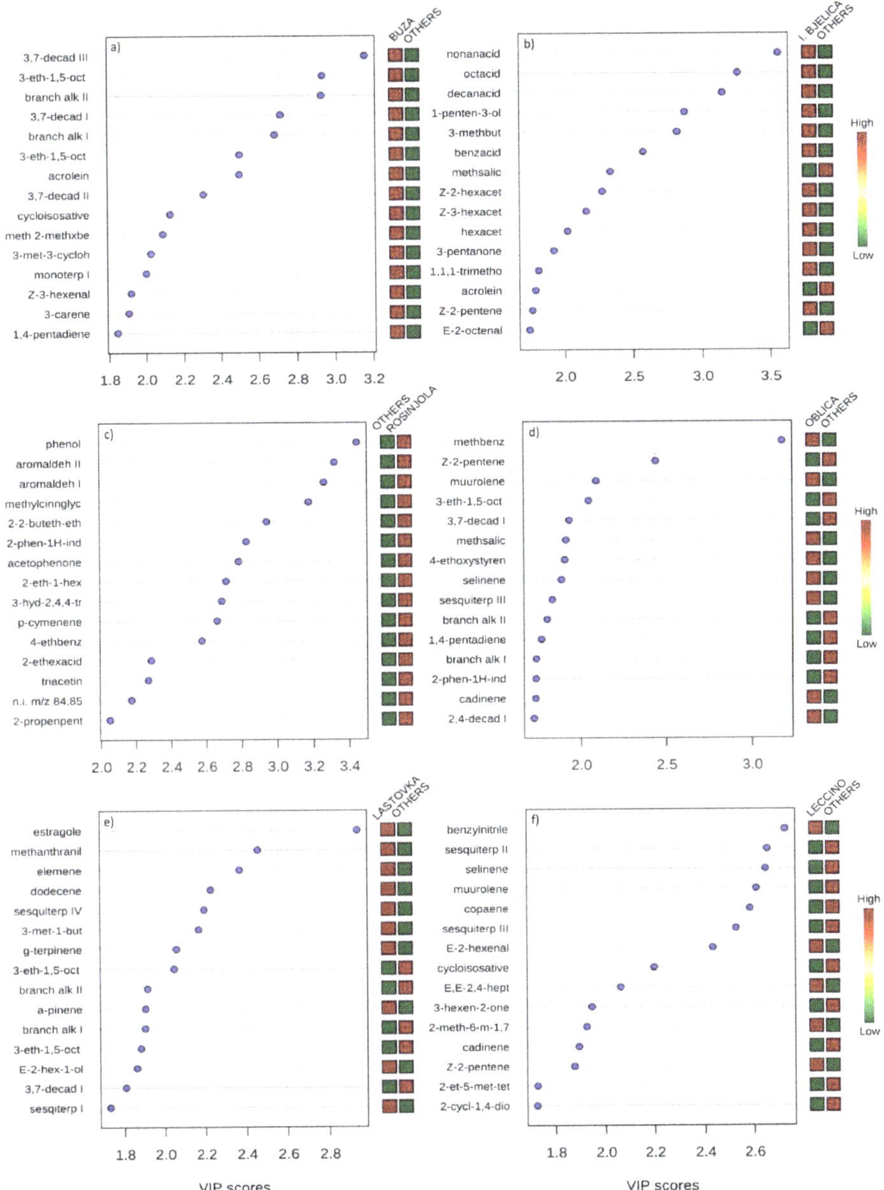

Figure 2. Variable Importance in Projection (VIP) scores of the variables (volatile compounds) most useful for the differentiation of each monovarietal extra virgin olive oil (EVOO), specifically: (**a**) Buža, (**b**) Istarska bjelica, (**c**) Rosinjola, (**d**) Oblica, (**e**) Lastovka, and (**f**) Leccino from the other five EVOOs produced in Croatia. Variables were extracted by partial least squares discriminant analysis applied on mean-centered data of six separate datasets each including two groups (a single vs. other five monovarietal EVOO).

3.2. Phenols

Nineteen phenolic compounds were identified in total, including simple phenols, phenolic acids, flavonoids, lignans, and secoiridoids (Table 3). For many of those significant differences between average concentrations in the investigated EVOOs were found, and a few phenols emerged as exclusive markers of particular monovarietal EVOOs. Lastovka EVOO was generally characterized by the highest concentrations of simple phenols (except vanillin) and *p*-coumaric acid. Leccino turned out to be clearly distinguishable from the other monovarietal EVOOs by the highest concentration of vanillin, and the lowest concentrations of *p*-coumaric acid and luteolin. Among lignans, high pinoresinol content was characteristic for Buža EVOO. The secoiridoid profiles differed among the investigated monovarietal EVOOs. The concentration of one of the major oleuropein aglycons and phenols in general in olive oil, dialdehydic form of decarboxymethylelenolic acid linked to hydroxytyrosol, i.e., 3,4-DHPEA-EDA or oleacein, was the highest in Leccino EVOO. Its tyrosol-based analogue, the major aglycon of ligstroside, *p*-HPEA-EDA or oleocanthal, clearly distinguished two groups of EVOO, I. bjelica, Oblica, and Leccino with higher, and Buža, Rosinjola, and Lastovka EVOOs with lower concentrations. The composition of other oleuropein and ligstroside aglycons also turned out to be variety-specific; it is worth mentioning low concentration of oleuropein aglycon I in Oblica and Leccino, low concentration of oleuropein aglycon II in Oblica, and exceptionally higher concentration of oleuropein + ligstroside aglycons in I. bjelica than in other EVOOs.

Table 3. Concentrations (mg/kg) of phenols determined by ultra-performance liquid chromatography with diode-array detection (UPLC-DAD) in monovarietal extra virgin olive oils produced from Buža, Istarska bjelica, Rosinjola, Oblica, Lastovka, and Leccino varieties in Croatia.

Phenol	Variety					
	Buža	I. bjelica	Rosinjola	Oblica	Lastovka	Leccino
simple phenols						
tyrosol	4.87 [b]	11.29 [a]	3.69 [b]	9.10 [ab]	12.28 [a]	5.60 [b]
hydroxytyrosol	5.40 [c]	10.21 [b]	5.59 [bc]	6.33 [bc]	20.17 [a]	6.47 [bc]
hydroxytyrosol acetate *	0.35 [c]	0.67 [b]	0.37 [bc]	0.42 [bc]	1.44 [a]	0.50 [bc]
vanillin	0.21 [b]	0.16 [bc]	0.20 [bc]	0.12 [c]	0.11 [c]	0.31 [a]
phenolic acids						
vanillic acid	0.31	0.32	0.33	0.18	0.37	0.25
p-coumaric acid	1.26 [bc]	0.90 [c]	0.82 [cd]	1.69 [b]	2.80 [a]	0.34 [d]
flavonoids						
luteolin	2.02 [bc]	2.95 [a]	2.86 [ab]	2.93 [a]	3.35 [a]	1.89 [c]
apigenin	0.55 [bc]	0.87 [a]	0.66 [b]	0.33 [d]	0.39 [d]	0.46 [cd]
lignans						
pinoresinol	9.97 [a]	4.02 [c]	6.98 [b]	3.21 [c]	3.68 [c]	4.14 [c]
Acetoxypinoresinol *	6.72 [c]	14.11 [a]	11.39 [ab]	8.69 [bc]	11.94 [ab]	7.49 [c]
secoiridoids						
Secologanoside *	0.03 [b]	0.04 [b]	0.03 [b]	0.04 [b]	0.06 [a]	0.04 [b]
elenolic acid glucoside *	0.04 [bc]	0.04 [c]	0.05 [abc]	0.05 [ab]	0.06 [a]	0.04 [c]
3,4-DHPEA-EDA *	95.50 [b]	115.68 [b]	104.93 [b]	98.46 [b]	121.33 [b]	175.06 [a]
oleuropein aglycone I *	72.56 [bc]	94.57 [ab]	109.85 [a]	49.09 [cd]	115.05 [a]	41.17 [d]
p-HPEA-EDA *	49.15 [b]	82.70 [a]	47.35 [b]	76.79 [a]	49.21 [b]	87.49 [a]
oleuropein + ligstroside aglycones I & II *	43.38 [b]	97.82 [a]	49.19 [b]	38.63 [b]	49.33 [b]	30.11 [b]
oleuropein aglycone II *	64.44 [c]	79.72 [abc]	100.61 [a]	42.38 [d]	94.14 [ab]	71.77 [bc]
ligstroside aglycon III *	1.66 [c]	4.60 [a]	1.82 [c]	2.79 [bc]	1.99 [c]	4.04 [ab]
oleuropein aglycone III *	9.06 [c]	15.84 [a]	11.66 [bc]	11.40 [bc]	13.75 [ab]	9.85 [c]
total phenols	367.25 [c]	536.49 [a]	458.38 [abc]	352.63 [c]	501.45 [ab]	447.00 [bc]

* The phenols for which pure standards were not available were quantified semi-quantitatively and their concentrations were expressed as equivalents of phenols with similar chemical structure assuming a response factor = 1. Different superscript lowercase letters in a row represent statistically significant differences between mean values at *p* < 0.05 obtained by one-way ANOVA and least significant difference (LSD) test.

3.3. Sensory Attributes

The majority of the investigated monovarietal EVOOs were characterized by common EVOO sensory attributes (Figure 3, Table S3). Buža EVOO showed higher intensities of the majority of the assessed positive odor attributes, and it was clearly distinguished from the others by the highest intensity of chicory/rocket. Istarska bjelica and Rosinjola had the lowest intensity of almond. The specificity of the odor of Oblica EVOO was contained mainly in the most intense green banana nuance, while Lastovka was distinguished as the only monovarietal EVOO with the woody note. Istrian EVOOs were generally described by higher intensities of green grass/leaves, aromatic herbs, and chicory/rocket attributes (with the exception of Leccino EVOO) in relation to the Dalmatian ones.

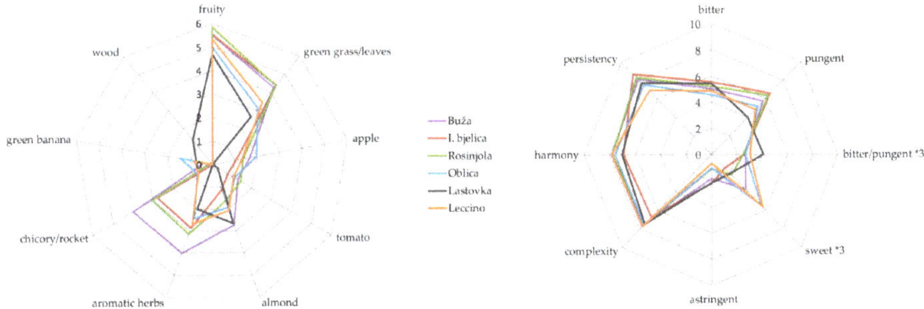

Figure 3. The intensities of the sensory attributes obtained by quantitative descriptive sensory analysis of monovarietal extra virgin olive oils (EVOOs) produced from Buža, Istarska bjelica, Rosinjola, Oblica, Lastovka, and Leccino varieties in Croatia (* 3 – the intensities of particular sensory attributes were multiplied by 3 to better visualize the differences between monovarietal EVOOs).

As regards the main EVOO taste attributes, bitterness and pungency, the EVOOs from Istrian native varieties Buža, I. bjelica, and Rosinjola generally showed higher intensities. The exception was the intensity of bitterness in Lastovka which was among the highest, which resulted in the highest bitterness/pungency ratio in this EVOO (Figure 3, Table S3). Oblica and Leccino were described as the sweetest among the EVOOs.

Istarska bjelica EVOO was characterized as the least complex, while I. bjelica and Lastovka EVOOs were less harmonious with respect to others. EVOOs from Istrian native varieties were the most, and Leccino the least persistent in terms of pungency.

4. Discussion

From the results of the GC-IT-MS and UPLC-DAD analysis of volatile compounds and phenols, respectively, it was clear that each of the investigated monovarietal EVOOs was characterized by a unique volatile and phenol profile. Since the samples were collected from various producers and were relatively heterogeneous in terms of geographical microlocations, growing conditions, harvest date, olive processing technology, and EVOO finalization and storage parameters, it could be assumed, with a high degree of certainty, that the effects of all of these factors were random, and that varietal origin was the main source of the observed differences. In fact, varietal origin was previously found to have a greater impact on volatile composition than various environmental factors [28]. However, the geographical origin possibly had an effect, which was impossible to evaluate separately from the effect of variety considering the varieties studied were specific for their regions. Istrian EVOOs, mostly those made of native Buža and Rosinjola, as well as those of the international variety Leccino, were characterized by higher concentrations of many LOX volatiles in relation to Dalmatian Oblica and Lastovka EVOO, including the most odoriferous ones, such as hexenals and 1-penten-3-one (Table 1). It is probable that this was directly reflected on the differences in their sensory profiles, since Istrian EVOOs had higher intensities of the majority of positive odor attributes, especially those of *green grass/leaves* and *chicory/rocket* (Figure 3, Table S3). Lower OAV values of the most potent odorants, (Z)-3-hexenal, 1-penten-3-one, and (E)-2-hexenal found in the Dalmatian, especially Lastovka EVOO, corroborated this assumption (Table 2). Since Istria is a region characterized by lower average temperatures than Dalmatia (Table S1), these results basically corroborated what was previously found in the majority of such studies that the temperature of environment is negatively correlated with the concentrations of LOX-derived volatile aroma compounds and the resulting EVOO positive sensory attributes [48,49].

Although without statistical significance in some cases, Buža EVOO excelled with the highest concentrations of the majority of positive LOX volatiles (Table 1), as well as with the highest intensities of positive odor sensory attributes (Figure 3, Table S3), which were probably in a causal relationship. It is worth emphasizing the highest cumulative odor activity value (OAV) of (Z)-3-hexenal and 1-penten-3-one, the two most powerful known odorants in EVOO with very low odor perception thresholds of 0.0017 and 0.00073 mg/kg, respectively [22,46] (Table 2), which certainly exhibited key roles. A large proportion of LOX volatiles among those extracted by the PLSDA as the most significant VIP compounds discriminating Buža from the other EVOOs (Figure 2) corroborated the assumption that this variety is characterized by strong lipoxygenase and hydroperoxide lyase activities in the LOX pathway.

Generally, the most similar to Buža in terms of high concentrations of LOX volatiles (Table 1) and their OAVs (Table 2), as well as high intensities of positive odor attributes (Figure 3, Table S3), was Rosinjola EVOO. When it came down to the discriminating VIP compounds, those extracted by PLSDA were mostly benzenoids (Figure 2). Many benzenoids which were found in relatively high concentration in Rosinjola EVOO, including methyl benzoate, acetophenone, and methyl salicylate, were previously reported to be important almond odorants [50], however their impact in olive oil has not been investigated yet. As well, *almond* note was not especially accentuated in Rosinjola EVOO (Figure 3, Table S3).

Istarska bjelica had lower concentrations of many important LOX volatiles (Table 1). Since it is a late ripening variety [51] it is possible that it was characterized by a slightly weaker LOX enzymatic load with respect to Buža and Rosinjola EVOO. As it is known that phenols may act as LOX enzymatic activity inhibitors [52], the possibility that the high concentrations of phenols found in this monovarietal EVOO (Table 3) acted in this way during milling and malaxation should not be excluded. However, the concentrations and OAVs of some other major LOX odorants, such as (E)-2-hexenal and 1-penten-3-one, were relatively high, suggesting a notable activity of (Z)-3:(E)-2-enal isomerase which catalyzes the conversion of (Z)-3- to (E)-2-hexenal, as well as relatively high activity of the enzymes or availability of the substrates involved in the synthesis of C5 compounds via 13-alkoxy radicals in this side-branch of the LOX pathway. The concentrations and OAVs of these volatiles were not lower that those found in Buža and Rosinjola EVOOs (Tables 1 and 2), so it is probable that (E)-2-hexenal and 1-penten-3-one were the key odorants in the formation of I. bjelica aroma and were the most responsible for the high intensity of several positive sensory attributes observed in this EVOO (Figure 3, Table S3). The most typical VIP chemical markers distinguishing I. bjelica EVOO were mostly non-LOX volatiles, namely middle-chain fatty acids and C6 alcohol acetates (Figure 2). Judging on the determined OAV values (Table 2), their sensory relevance was probably minor to medium. High concentrations of C6 alcohol acetates (Table 1) implied a possible high alcohol acyl transferase activity in olives and olive paste of this variety [53].

As stated previously, Oblica was characterized by a slightly lower contribution of the LOX volatiles and, consequently, lower intensities of particular positive odor attributes with respect to Istrian EVOOs, but was still superior to Lastovka EVOO (Table 1, Figure 3, Table S3). It was possibly mostly due to lower 1-penten-3-one and (E)-2-hexenal concentrations and OAVs, since the level of (Z)-3-hexenal was relatively high (Tables 1 and 2). As well, it is possible that a part of the *fruity* and *green* aroma originated from hexanal, found in higher concentration with respect to the Istrian EVOOs (Tables 1 and 2). *Green banana* odor sensory attribute which was found to be typical for Oblica EVOO (Figure 3, Table S3) could have, at least partly, originated from the volatiles often associated with this nuance. (Z)-3-hexen-1-ol was certainly a candidate for this role [45], since its concentration was the highest in this EVOO (Table 1) and at the same time above the corresponding odor detection threshold (Table 2). For other LOX volatiles with the odor commonly described as banana-like, such as hexanol, hexenyl acetates and penten-1-ols [45,54], no significant differences between varieties were found. As well, their levels in Oblica were not among the highest among the investigated EVOOs (Table 1), implying their impact in the formation of *green banana* nuance was probably not crucial. The same applies for other minor volatiles commonly reported as carriers of banana odor, such as isoamyl and other acetates. The VIP compounds responsible for the differentiation of Oblica EVOO (Figure 2), which pertained to several chemical families, could have not been meaningfully related to the occurrence of *green banana* odor.

Lastovka EVOO was characterized by the most distinguishable volatile profile among the investigated monovarietal EVOOs. It contained the lowest concentrations of the majority of LOX volatiles (Table 1), including the most potent odorants with the highest OAVs (Table 2), which was certainly a direct cause of the lowest intensities of the majority of positive odor attributes perceived in this EVOO (Figure 3, Table S3). On the other hand, it was found to have high amounts of hexanal, particular C6 alcohols, and hexyl acetate, compounds often accounted among the carriers of *green* odor which derive from the enzymatic degradation of linoleic acid but also oxidation [10,28,55]. Lastovka EVOO contained the highest concentrations of particular monoterpenes and sesquiterpenes, whose sensory contribution is generally described by descriptors such as *citrus, camphor, eucalyptus, roses,* etc., as well as *wood*. Although sensory relevance of terpenes in olive oil is currently still unknown and it is certainly limited by the lipid matrix in which these lipophilic molecules are highly soluble, the possibility of their contribution to the specific *wood* odor perceived in Lastovka EVOO during sensory analysis (Figure 3, Table S3) should not be excluded. Several terpenes were extracted by PLSDA as among the most discriminative compounds for this variety (Figure 2). Particular sesquiterpenes, on the other hand, such as (+)-cycloisosativene, α-copaene, α-muurolene, δ-cadinene, and several unidentified ones were found in the lowest concentrations in Lastovka EVOO (Table 1). Terpenes were previously found to have large potential to differentiate EVOO according to variety [16,56], which was basically confirmed in this study. Other compounds found to be characteristic for Lastovka EVOO, such as particular saturated short-chain aldehydes, ketones, alcohols, and acids, could also have had a sensory impact with their *malty, pungent, rancid,* and *sweaty* nuances (Table 2).

One of probably the most important characteristics found typical for Leccino EVOO was the ratio between the important (E)-2- and (Z)-3-C6 forms, which was generally the highest and discriminated well this EVOO from the majority of the other studied EVOOs (Tables 1 and 2). The highest concentration of (E)-2-hexenal and the lowest concentration of (Z)-3-hexenal, (Z)-3-hexen-1-ol, as well as the low concentration of (Z)-3-hexenyl acetate in Leccino EVOO were likely the result of high (Z)-3:(E)-2-enal isomerase activity in Leccino olives, i.e., olive paste during milling and malaxation steps [57]. Considering that the estimated contribution of (E)-2-hexenal to the aroma of EVOOs was generally lower than that of (Z)-3-hexenal (Table 2), it is possible that one of the consequences of the observed differences was a slightly lower intensity of particular odor sensory attributes, such as *green/grass leaves* and *chicory/rocket*, observed in Leccino with respect to the EVOOs from the other, native Istrian varieties Buža, I. bjelica and Rosinjola (Figure 3, Table S3). Other interesting features of Leccino EVOO included lower concentrations of particular sesquiterpenes and furanoids. In fact, many sesquiterpenes were among those with the highest VIP scores extracted by PLSDA, but with a negative sign (Figure 2).

Phenols, especially secoiridoids, are responsible for the characteristic EVOO bitterness and pungency, but the specific sensory contribution of each individual major secoiridoid has not been precisely elucidated up to date. Nevertheless, there is solid evidence that *p*-HPEA-EDA is a key contributor to pungency, while the pungency of other, monoaldehydic ligstroside aglycons is weaker, although still strong [24]. Ligstroside aglycons were found to generally be less bitter than pungent, which was especially the case for *p*-HPEA-EDA. In the same study [24] it was found that the majority of oleuropein aglycons, including 3,4-DHPEA-EDA, was described as both bitter and pungent, with some of them exhibiting rather strong bitterness. The lowest intensity of bitterness observed in Oblica and Leccino (Figure 2, Table S3) could be tentatively linked to the lowest oleuropein aglycon I concentrations found in these EVOO (Table 3). The intensity of pungency did not quite correlate with the average *p*-HPEA-EDA concentrations (Figure 3, Table 3, Table S3). In contrast to I. bjelica, for which a positive correlation was observed, Rosinjola EVOO was characterized as intensively pungent according to the official method [1] (intensity >6) despite containing relatively low concentration of this secoiridoid. The highest average concentration of oleuropein aglycon II in Rosinjola EVOO (Table 3) could have possibly compensated for this deficiency. The pungency of I. bjelica possibly partly originated also from the highest concentrations of all the three monoaldehydic ligstroside aglycons found in this EVOO (Table 3). Especially interesting was the highest ratio of bitterness to pungency found in Lastovka EVOO (Figure 3, Table S3). Roughly, Lastovka EVOO contained among the highest concentrations of oleuropein aglycones and among the lowest concentrations of ligstroside aglycones, which could have had such an impact. This EVOO had the highest concentration of *p*-coumaric acid which, although relatively low, possibly contributed to the bitterness observed. It is worth mentioning that the UPLC chromatograms of Lastovka EVOO contained several unidentified peaks in addition to those observed in the other monovarietal EVOOs (data not shown), which possibly originated from the compounds with sensory relevance. The so-called *sweetness* in most cases coincided with the lower amounts of total phenols, which was as expected (Figure 3, Table 3). Again, several features turned out to be specific for Leccino EVOO (Table 3), the most important being the highest concentration of 3,4-DHPEA-EDA, which implied variety-dependent differences with respect to the availability of precursors and enzymatic activity between Leccino and Croatian native olive varieties.

5. Conclusions

The use of GC-IT-MS and UPLC-DAD proved to be a powerful combination for studying the inter-varietal diversity of typical volatile and phenolic profiles of Croatian EVOOs, respectively. Each of the investigated monovarietal EVOO displayed unique volatile aroma and phenol composition. The qualitative and quantitative chromatographic data was useful for tentative elucidation of some of the perceived sensory attributes including the variety-typical ones, which though has to be taken with caution due to the extreme complexity of the established chemical profiles with the majority of volatiles with still unknown sensory relevance. Many potential varietal markers were extracted by uni- and multivariate statistical analysis despite high intra-varietal heterogeneity. It was demonstrated that volatiles and phenols from all the investigated chemical classes can be useful for this purpose. Many of the volatile compounds which turned out to have a notable discrimination power were (tentatively) identified for the first time in EVOO, or were generally neglected in previous studies, especially from sensorial point of view. In fact, only in a few cases were the major LOX compounds, studied most extensively among the volatiles up to date, sufficient for a robust varietal differentiation in this work. This indicates a large potential of the untargeted fingerprinting approach for EVOO characterization, differentiation, and authentication studies.

The number of the extracted robust varietal markers among the investigated chemical compounds largely exceeded the number of typical sensory attributes useful to differentiate monovarietal EVOOs. It is reasonable to conclude that the approach which comprises GC-IT-MS and UPLC-DAD analytical techniques may provide additional objective information about varietal origin which successfully complement those obtained by sensory analysis. Probably the best example for this is the case of Rosinjola EVOO which was relatively similar and hardly distinguishable from that of Buža variety based solely on the sensory analysis, but was characterized by many exclusive chemical markers among benzenoid and furanoid volatiles which discriminated this EVOO rather successfully.

The results obtained in this study could certainly be useful for improving the quality management and control in the production of Croatian monovarietal/PDO EVOO. These findings could contribute to strengthening their PDO identities and position on the market, and could be especially useful for discriminating EVOOs of Croatian native varieties from the world famous Leccino variety.

Supplementary Materials: The following are available online at http://www.mdpi.com/2304-8158/8/11/565/s1, Table S1: Climate parameters in the Istria and Dalmatia regions of Croatia in 2015, Table S2: Standardized coefficients of the variables selected for the differentiation of monovarietal Buža, Istarska bjelica, Rosinjola, Oblica, Lastovka, and Leccino extra virgin olive oils on the first three discriminant functions obtained by stepwise linear discriminant analysis, and the percentage of correct classification at each step, Table S3: The intensities and scores of the sensory attributes perceived in monovarietal extra virgin olive oils produced from Buža, Istarska bjelica, Rosinjola, Oblica, Lastovka, and Leccino varieties in Croatia.

Author Contributions: Conceptualization, I.L.; Methodology, I.L.; Formal analysis, I.L., M.L., M.Ž., M.K., S.G., K.B.B.; Resources, I.L., M.L., K.B.B.; Data curation, I.L., M.L., K.B.B.; Writing—original draft preparation, I.L.; Writing—review and editing, I.L., M.L., K.B.B.; Supervision, I.L.; Project administration, I.L.; Funding acquisition, I.L.

Funding: This research was funded by Croatian Science Foundation grant number UIP-2014-09-1194.

Acknowledgments: The authors would like to thank olive oil producers from Istria and Dalmatia (Croatia, EU) for donating the samples of extra virgin olive oils, and Ivana Horvat M.Sc. for technical assistance.

Conflicts of Interest: The authors declare no conflict of interest. The funders had no role in the design of the study; in the collection, analyses, or interpretation of data; in the writing of the manuscript, or in the decision to publish the results.

References

1. European Economic Community. Commission Regulation (EEC) No 2568/91 of 11 July 1991 (and later modifications) on the characteristics of olive oil and olive-residue oil and the relevant methods of analysis. *Off. J. Eur. Communities* **1991**, *L248*, 1–83.

2. European Union. Regulation (EU) No 1151/2012 of the European parliament and of the Council of 21 November 2012 (and later modifications) on quality schemes for agricultural products and foodstuffs. *Off. J. Eur. Union* **2012**, *L343*, 1–29.

3. International Olive Council. *COI/T.20/Doc. No. 22 Method for the Organoleptic Assessment of Extra Virgin Olive Oil Applying to Use a Designation of Origin*; International Olive Oil Council: Madrid, Spain, 2005.

4. Daisa, P.; Hatzakis, E. Quality assessment and authentication of virgin olive oil by NMR spectroscopy: A critical review. *Anal. Chim. Acta* **2013**, *765*, 1–27. [CrossRef] [PubMed]

5. Aparicio, R.; Morales, M.T.; Aparicio-Ruiz, R.; Tena, N.; García-González, D.L. Authenticity of olive oil: Mapping and comparing official methods and promising alternatives. *Food Res. Int.* **2013**, *54*, 2025–2038. [CrossRef]

6. López-Cortés, I.; Salazar-García, D.C.; Velázquez-Martí, B.; Salazar, D.M. Chemical characterization of traditional varietal olive oils in East of Spain. *Food Res. Int.* **2013**, *54*, 1934–1940. [CrossRef]

7. Olmo-García, L.; Polari, J.J.; Li, X.; Bajoub, A.; Fernández-Gutiérrez, A.; Wang, S.C.; Carrasco-Pancorbo, A. Study of the minor fraction of virgin olive oil by a multi-class GC–MS approach: Comprehensive quantitative characterization and varietal discrimination potential. *Food Res. Int.* **2019**, *125*, 108649. [CrossRef]

8. Kritioti, A.; Menexes, G.; Drouza, C. Chemometric characterization of virgin olive oils of the two major Cypriot cultivars based on their fatty acid composition. *Food Res. Int.* **2018**, *103*, 426–437. [CrossRef]

9. Angerosa, F. Virgin olive oil odour notes: Their relationships with volatile compounds from the lipoxygenase pathway and secoiridoid compounds. *Food Chem.* **2000**, *68*, 283–287. [CrossRef]

10. Angerosa, F. Influence of volatile compounds on virgin olive oil quality evaluated by analytical approaches and sensor panels. *Eur. J. Lipid Sci. Technol.* **2002**, *104*, 639–660. [CrossRef]

11. Angerosa, F.; Campestre, C. Sensory quality: Methodologies and applications. In *Handbook of Olive Oil—Analysis and Properties*; Aparicio, R., Harwood, J., Eds.; Springer: New York, NY, USA, 2013; pp. 523–560.

12. Campestre, C.; Angelini, G.; Gasbarri, C.; Angerosa, F. The compounds responsible for the sensory profile in monovarietal virgin olive oils. *Molecules* **2017**, *22*, 1833. [CrossRef]

13. Servili, M.; Montedoro, G. Contribution of phenolic compounds to virgin olive oil quality. *Eur. J. Lipid Sci. Technol.* **2002**, *104*, 602–613. [CrossRef]

14. Taticchi, A.; Esposto, S.; Servili, M. The basis of the sensory properties of virgin olive oil. In *Olive Oil Sensory Science*; Monteleone, E., Langstaff, S., Eds.; John Wiley & Sons, Ltd.: Hoboken, NJ, USA, 2014; pp. 33–54.

15. Blasi, F.; Pollini, L.; Cossignani, L. Varietal authentication of extra virgin olive oils by triacylglycerols and volatiles analysis. *Foods* **2019**, *8*, 58. [CrossRef] [PubMed]

16. Cecchi, T.; Alfei, B. Volatile profiles of Italian monovarietal extra virgin olive oils via HS-SPME–GC–MS: Newly identified compounds, flavors molecular markers, and terpenic profile. *Food Chem.* **2013**, *141*, 2025–2035. [CrossRef] [PubMed]

17. García-González, D.L.; Romero, N.; Aparicio, R. Comparative study of virgin olive oil quality from single varieties cultivated in Chile and Spain. *J. Agric. Food Chem.* **2010**, *58*, 12899–12905. [CrossRef] [PubMed]

18. Ocakoglu, D.; Tokatli, F.; Ozen, B.; Korel, F. Distribution of simple phenols, phenolic acids and flavonoids in Turkish monovarietal extra virgin olive oils for two harvest years. *Food Chem.* **2009**, *113*, 401–410. [CrossRef]

19. Piscopo, A.; De Bruno, A.; Zappia, A.; Ventre, C.; Poiana, M. Characterization of monovarietal olive oils obtained from mills of Calabria region (Southern Italy). *Food Chem.* **2016**, *213*, 313–318. [CrossRef] [PubMed]

20. Sagratini, G.; Maggi, F.; Caprioli, G.; Cristalli, G.; Ricciutelli, M.; Torregiani, E.; Vittori, S. Comparative study of aroma profile and phenolic content of Montepulciano monovarietal red wines from the Marches and Abruzzo regions of Italy using HS-SPME–GC–MS and HPLC–MS. *Food Chem.* **2012**, *132*, 1592–1599. [CrossRef] [PubMed]

21. Angerosa, F.; Servili, M.; Selvaggini, R.; Taticchi, A.; Esposto, S.; Montedoro, G. Volatile compounds in virgin olive oil: Occurrence and their relationship with the quality. *J. Chromatogr. A* **2004**, *1054*, 17–31. [CrossRef]

22. Kalua, C.M.; Allen, M.S.; Bedgood, D.R.; Bishop, A.G.; Prenzler, P.D.; Robards, K. Olive oil volatile compounds, flavour development and quality: A critical review. *Food Chem.* **2007**, *100*, 273–286. [CrossRef]

23. Vichi, S.; Pizzale, L.; Conte, L.S.; Buxaderas, S.; López-Tamames, E. Solid-Phase microextraction in the analysis of virgin olive oil volatile fraction: Characterization of virgin olive oils from two distinct geographical areas of Northern Italy. *J. Agric. Food Chem.* **2003**, *51*, 6572–6577. [CrossRef]

24. Andrewes, P.; Busch, J.L.H.C.; De Joode, T.; Groenewegen, A.; Alexandre, H. Sensory properties of virgin olive oil polyphenols: Identification of deacetoxy-ligstroside aglycon as a key contributor to pungency. *J. Agric. Food Chem.* **2003**, *51*, 1415–1420. [CrossRef] [PubMed]

25. Bendini, A.; Cerretani, L.; Carrasco-Pancorbo, A.; Gómez-Caravaca, A.M.; Segura-Carretero, A.; Fernández-Gutiérrez, A.; Lercker, G. Phenolic molecules in virgin olive oils: A survey of their sensory properties, health effects, antioxidant activity and analytical methods. An overview of the last decade. *Molecules* **2007**, *12*, 1679–1719. [CrossRef] [PubMed]

26. Amirante, P.; Clodoveo, M.L.; Tamborrino, A.; Leone, A.; Dugo, G. Oxygen concentration control during olive oil extraction process: A new system to emphasize the organoleptic and healthy properties of virgin olive oil. *Acta Hortic.* **2012**, *949*, 473–480. [CrossRef]

27. Taticchi, A.; Esposto, S.; Veneziani, G.; Urbani, S.; Selvaggini, R.; Servili, M. The influence of the malaxation temperature on the activity of polyphenoloxidase and peroxidase and on the phenolic composition of virgin olive oil. *Food Chem.* **2013**, *136*, 975–983. [CrossRef] [PubMed]

28. Angerosa, F.; Basti, C.; Vito, R. Virgin olive oil volatile compounds from lipoxygenase pathway and characterization of some Italian cultivars. *J. Agric. Food Chem.* **1999**, *47*, 836–839. [CrossRef] [PubMed]

29. Cirilli, M.; Caruso, G.; Gennai, C.; Urbani, S.; Frioni, E.; Ruzzi, M.; Servili, M.; Gucci, R.; Poerio, E.; Muleo, R. The role of polyphenoloxidase, peroxidase, and β-glucosidase in phenolics accumulation in *Olea europaea* L. fruits under different water regimes. *Front. Plant Sci.* **2017**, *8*, 717. [CrossRef] [PubMed]

30. Ministry of Agriculture, Republic of Croatia. Available online: https://poljoprivreda.gov.hr/istaknute-teme/hrana-111/oznake-kvalitete/zoi-zozp-zts-poljoprivrednih-i-prehrambenih-proizvoda/zasticena-oznaka-izvornosti-zoi/1206 (accessed on 9 October 2019).

31. International Olive Council, Madrid, Spain. Available online: http://www.internationaloliveoil.org/estaticos/view/131-world-olive-oil-figures (accessed on 9 October 2019).

32. Poljuha, D.; Sladonja, B.; Šetić, E.; Milotić, A.; Bandelj, D.; Jakše, J.; Javornik, B. DNA fingerprinting of olive varieties in Istria (Croatia) by microsatellite markers. *Sci. Hortic.* **2008**, *115*, 223–230. [CrossRef]

33. Brkić Bubola, K.; Koprivnjak, O.; Sladonja, B.; Lukić, I. Volatile compounds and sensory profiles of monovarietal virgin olive oils from Buža, Črna and Rosinjola cultivars in Istria (Croatia). *Food Technol. Biotechnol.* **2012**, *50*, 192–198.

34. Giacometti, J.; Milin, Č.; Giacometti, F.; Ciganj, Z. Characterisation of monovarietal olive oils obtained from Croatian cvs. Drobnica and Buža during the ripening period. *Foods* **2018**, *7*, 188. [CrossRef]

35. Kulišić Bilušić, T.; Melliou, E.; Giacometti, J.; Čaušević, A.; Čorbo, S.; Landeka, M.; Magiatis, P. Phenolics, fatty acids, and biological potential of selected Croatian EVOOs. *Eur. J. Lipid Sci. Technol.* **2017**, *119*, 1700108. [CrossRef]

36. Lukić, I.; Carlin, S.; Horvat, I.; Vrhovsek, U. Combined targeted and untargeted profiling of volatile aroma compounds with comprehensive two-dimensional gas chromatography for differentiation of virgin olive oils according to variety and geographical origin. *Food Chem.* **2019**, *270*, 403–414. [CrossRef] [PubMed]

37. Lukić, I.; Krapac, M.; Lukić, M.; Vrhovsek, U.; Godena, S.; Brkić Bubola, K. Towards understanding the varietal typicity of virgin olive oil by correlating sensory and compositional analysis data: A case study. *Food Res. Int.* **2018**, *112*, 78–89. [CrossRef] [PubMed]

38. Šarolić, M.; Gugić, M.; Friganović, E.; Tuberoso, C.I.G.; Jerković, I. Phytochemicals and other characteristics of Croatian monovarietal extra virgin olive oils from Oblica, Lastovka and Levantinka varieties. *Molecules* **2015**, *20*, 4395–4409. [CrossRef] [PubMed]

39. Žanetić, M.; Cerretani, L.; Škevin, D.; Politeo, O.; Vitanović, E.; Jukić Špika, M.; Perica, S.; Ožić, M. Influence of polyphenolic compounds on the oxidative stability of virgin olive oils from selected autochthonous varieties. *J. Food Agric. Environ.* **2013**, *11*, 126–131.

40. Žanetić, M.; Škevin, D.; Vitanović, E.; Jukić Špika, M.; Perica, S. Survey of phenolic compounds and sensorial profile of Dalmatian virgin olive oils. *Pomol. Croat.* **2011**, *17*, 19–30.

41. Brkić Bubola, K.; Koprivnjak, O.; Sladonja, B.; Škevin, D.; Belobrajić, I. Chemical and sensorial changes of Croatian monovarietal olive oils during ripening. *Eur. J. Lipid Sci. Technol.* **2012**, *114*, 1400–1408. [CrossRef]

42. Jerman Klen, T.; Golc Wondra, A.; Vrhovšek, U.; Mozetič Vodopivec, B. Phenolic profiling of olives and olive oil process-derived matrices using UPLC-DAD-ESI-QTOF-HRMS analysis. *J. Agric. Food Chem.* **2015**, *63*, 3859–3872. [CrossRef]

43. Xia, J.; Sinelnikov, I.; Han, B.; Wishart, D.S. MetaboAnalyst 3.0—Making metabolomics more meaningful. *Nucleic Acids Res.* **2015**, *43*, W251–W257. [CrossRef]

44. Aparicio, R.; Luna, G. Characterization of monovarietal virgin olive oils. *Eur. J. Lipid Sci. Technol.* **2002**, *104*, 614–627. [CrossRef]

45. Morales, M.T.; Luna, G.; Aparicio, R. Comparative study of virgin olive oil sensory defects. *Food Chem.* **2005**, *91*, 293–301. [CrossRef]

46. Reiners, J.; Grosch, W. Odorants of virgin olive oils with different flavor profiles. *J. Agric. Food Chem.* **1998**, *46*, 2754–2763. [CrossRef]

47. Reboredo-Rodríguez, P.; González-Barreiro, C.; Cancho-Grande, B.; Simal-Gándara, J. Improvements in the malaxation process to enhance the aroma quality of extra virgin olive oils. *Food Chem.* **2014**, *158*, 534–545. [CrossRef] [PubMed]

48. Inglese, P.; Famiani, F.; Galvano, F.; Servili, M.; Esposto, S.; Urbani, S. Factors affecting extra-virgin olive oil composition. In *Horticultural Reviews*; Janick, J., Ed.; John Wiley & Sons Inc.: Hoboken, NJ, USA, 2010; pp. 83–147.

49. Romero, N.; Saavedra, J.; Tapia, F.; Sepúlveda, B.; Aparicio, R. Influence of agroclimatic parameters on phenolic and volatile compounds of Chilean virgin olive oils and characterization based on geographical origin, cultivar and ripening stage. *J. Sci. Food Agric.* **2016**, *96*, 583–592. [CrossRef] [PubMed]

50. Beck, J.J.; Higbee, B.S.; Light, D.M.; Gee, W.S.; Merrill, G.B.; Hayashi, J.M. Hull split and damaged almond volatiles attract male and female navel orangeworm moths. *J. Agric. Food Chem.* **2012**, *60*, 8090–8096. [CrossRef] [PubMed]

51. Lukić, I.; Krapac, M.; Horvat, I.; Godena, S.; Kosić, U.; Brkić Bubola, K. Three-factor approach for balancing the concentrations of phenols and volatiles in virgin olive oil from a late-ripening olive cultivar. *LWT Food Sci. Technol.* **2018**, *87*, 194–202. [CrossRef]

52. Majetić Germek, V.; Koprivnjak, O.; Butinar, B.; Pizzale, L.; Bučar-Miklavčič, M.; Conte, L.S. Influence of phenols mass fraction in olive (*Olea europaea* L.) paste on volatile compounds in Buža cultivar virgin olive oil. *J. Agric. Food Chem.* **2013**, *61*, 5921–5927.

53. Angerosa, F.; Mostallino, R.; Basti, C.; Vito, R. Influence of malaxation temperature and time on the quality of virgin olive oils. *Food Chem.* **2001**, *72*, 19–28. [CrossRef]

54. Aparicio, R.; Morales, M.T. Characterization of olive ripeness by green aroma compounds of virgin olive oil. *J. Agric. Food Chem.* **1998**, *46*, 1116–1122. [CrossRef]

55. Hongsoongnern, P.; Chambers, E. A lexicon for green odor or flavor and characteristics of chemicals associated with green. *J. Sens. Stud.* **2008**, *23*, 205–221. [CrossRef]
56. Vichi, S.; Guadayol, J.M.; Caixach, J.; Lopez-Tamames, E.; Buxaderas, S. Monoterpene and sesquiterpene hydrocarbons of virgin olive oil by headspace solidphase microextraction coupled to gas chromatography/mass spectrometry. *J. Chromatogr. A* **2006**, *1125*, 117–123. [CrossRef]
57. Williams, M.; Salas, J.J.; Sanchez, J.; Harwood, J.L. Lipoxygenase pathway in olive callus cultures (*Olea europaea*). *Phytochemistry* **2000**, *53*, 13–19. [CrossRef]

 *foods*

Article

Extra Virgin Olive Oil Quality as Affected by Yeast Species Occurring in the Extraction Process

Simona Guerrini [1] , Eleonora Mari [2], Damiano Barbato [2] and Lisa Granchi [2,*]

[1] FoodMicroTeam s.r.l., Academic Spin-Off of the University of Florence, via Santo Spirito, 14-50125 Florence, Italy; simona@foodmicroteam.it

[2] Department of Agriculture, Food, Environment and Forestry (DAGRI), P.le delle Cascine, 24-50144 Florence, Italy; eleonora.mari@unifi.it (E.M.); damiano.barbato@unifi.it (D.B.)

* Correspondence: lisa.granchi@unifi.it; Tel.: +39-055-275-5916

Received: 6 September 2019; Accepted: 5 October 2019; Published: 7 October 2019

Abstract: In extra virgin olive oil (EVOO) extraction process, the occurrence of yeasts that could affect the quality of olive oil was demonstrated. Therefore, in this work, at first, the yeasts occurring during different extractive processes carried out in a Tuscany oil mill, at the beginning, in the middle, and the end of the harvesting in the same crop season, were quantified. Then, possible effects on quality of EVOO caused by the predominant yeast species, possessing specific enzymatic activities, were evaluated. Yeast concentrations were higher in extraction processes at the end of the harvesting. Twelve yeast species showing different isolation frequencies during olive oil extractive process and according to the harvesting date were identified by molecular methods. The yeast species dominating olive oil samples from decanter displayed enzymatic activities, potentially affecting EVOO quality according to zymogram analysis. HS-SPME-GC-MS analysis of the volatile compounds in commercial EVOO, inoculated with three yeast species (*Nakazawaea molendini-olei*, *Nakazawaea wickerhamii*, *Yamadazyma terventina*), pointed out significant differences depending on the strain inoculated. In conclusion, during the olive oil extractive processes, some yeast species colonize the extraction plant and may influence the chemical and sensory characteristics of EVOO depending on the cell concentrations and their enzymatic capabilities.

Keywords: yeast microbiota; extra virgin olive oil; *Nakazawaea molendini-olei*; *Nakazawaea wickerhamii*; *Yamadazyma terventina*; yeast enzymatic activities; volatile compounds; sensory analysis

---

1. Introduction

Extra virgin olive oil (EVOO) is not just a product obtained from the fruit of the olive tree by mechanical extraction, but rather the result of complex changes in fruit components. Because of these changes, chemical compounds affecting the qualitative characteristics for sensory acceptability of extra virgin olive oil [1] may be produced. Pleasant sensory notes, characterizing extra virgin olive oil, are mainly originated from aldehydes, esters, alcohols, and ketones, which are responsible for oil sensory attributes such as "green" and "fruity" [2–6]. Nevertheless, several phenomena can alter the initial pleasant flavor, giving rise to unpleasant sensory notes, classified, according to the current olive oil regulations (EU Reg. 1348/2013), into four groups: "fusty", "musty", "winey–vinegary", and "rancid". Microorganisms associated with the olives may affect oil quality according to their metabolic activities. Indeed, as reported by Vichi et al. [7–9], oils from microbiologically contaminated olives exhibited a lower quality level and influences of olive microbiota on oil characteristics were greater than the effects exerted by malaxation time and temperature. Guerrini et al. [10] showed that sensory defects and specific volatile compounds (i.e., 2-butanone, butyric acid, 2-heptanol, octanoic acid, 1-octen-3-ol) were correlated to both yeast and mould concentrations detected in extracted and filtered oils. Yeasts and moulds are present in extracted oil because, during olive crushing, microorganisms of olives pass on into oil through both solid particles of olive fruit and micro-drops of vegetation water [11–13]. Some yeast species occurring in newly unfiltered oil can remain viable and metabolically active during the conservation period and, according to their metabolic capabilities, can either improve or worsen the oil quality [9]. Enzymatic activities of yeasts isolated from either olives or olive oil have been reported to include β-glucosidase, β-glucanase, polyphenoloxidases, peroxidase, lipase and cellulase activities [11,14–17]. Enzymes such as ß-glucosidase are known to improve oil quality by increasing phenolic compound extractability, while others such as lipase, polyphenoloxidases, and peroxidase are known to cause detrimental effects [5,18–20]. Recent studies [17,21] demonstrated that the presence of some yeast species might be responsible for olive oil sensory decay during storage. In particular, laboratory experiments showed the presence of defects in olive oil treated with specific yeast strains of *Candida adriatica*, *Nakazawaea wickerhamii*, and *Candida diddensiae*, while other olive oil samples treated with other *Candida diddensiae* strains were defect-free after four months of storage [22,23]. By the way, yeasts belonging to various genera were isolated in commercial extra virgin olive oil (*Candida*, *Nakazawaea*, *Williopsis*, *Ogataea*, *Yamadazyma* and *Saccharomyces*) [11–14,24–27].

Despite these evidence regarding the presence of viable yeasts in oil and their potential impact on olive oil quality, only few studies have investigated the yeast species occurring in the different phases of the olive oil extraction process and their effects on the oil quality [16,23,28]. In particular, a recent study of Mari et al. [28]. Mari et al. showed that the yeast populations occurring in olive oil extraction processes are numerically significant and originate principally from the yeasts colonizing the oil extractive plants. In fact, this study showed that only three of the eleven dominant yeast species detected on the washed olives were also found in extracted oil at significant isolation frequencies (*Candida adriatica*, *Nakazawaea molendini-olei*, and *Nakazawaea wickerhamii*). On the contrary, some yeast species showed significant isolation frequencies only in extracted oil (*Yamadazyma terventina*), or in kneaded pastes and pomaces (*Zygotorulaspora mrakii*). The occurrence of different yeast species according to the source of isolation (pastes, extracted oil or pomaces) suggests a contamination of the plant during oil extraction that select specific yeast species [28]. Ciafardini et al. [29] found a lower species diversity based on the origin of isolation. These Authors found six different yeast species (*Kluyveromyces marxianus*, *Candida oleophila*, *Candida diddensiae*, *Candida norvegica*, *Wickerhamomyces anomalus* and *Debaryomyces hansenii*). Except from *K. marxianus* that was found only in the wash water and *W. anomalus* that was found only in the six-month stored olive oil, all the other species occurred in the wash water and in the kneaded paste as well as in the newly produced olive oil. Anyway, a selected microbiota, when numerically significant, could affect olive oil quality in different ways, based on the specific metabolic capabilities of each yeast species or even strain. Therefore, the aim of this study was to assess whether the yeast microbiota occurring in olive oil extraction process affects the quality of extra virgin olive oil and, in particular, the volatile compounds content

For this purpose, at first, yeast species present during different extractive processes carried out in the same crop season as well as the chemical and sensory characteristics of the resulting olive oils were investigated. Then, some isolates belonging to the yeast species present at higher frequency in the process and possessing some enzymatic activities were inoculated into a commercial olive oil in order to assess their effective effects on extra-virgin olive oil quality.

2. Materials and Methods

2.1. Sampling throughout Olive Oil Extraction Processes

During the same crop season, 14 batches of approx. 200 kg olives, form Frantoio, Moraiolo, and mixed cultivars, were processed in a Tuscany oil mill (Azienda Agricola Buonamici s.r.l., Fiesole, Florence, Italy). Olives were collected and processed, within 4 h of harvesting, in three different harvest time at ten-day intervals: 6 at the beginning (HD1a, HD1b, HD1c, HD1d, HD1e and HD1f), 5 in the middle (HD2a, HD2b, HD2c, HD2d and HD2e) and 3 at the end (HD3a, HD3b and HD3c). Plant for oil extraction (TEM, Florence, Italy) consisted of a cleaning and water washing system, an olive grinding cutter crusher (mod. FR350), a controlled-temperature vertical axis malaxation equipment (500 kg capacity) (mod. V500), a "decanter" (two-step mod. D1500) with 1500 kg/h maximum capacity and a cardboard filter press (15 μm cut-off). Plastic residue or "alperujo" from decanter was subjected to separation by centrifugation of stone fragments to obtain destoned pomace. Olives were crushed at 2500 rpm (crusher holes 6.5 mm in diameter); malaxation was carried out at half capacity under vacuum (residual pressure of 20 kPa) at 22 ± 1 °C for a mean time of 15 min. Decanter worked with a screw conveyor rotating at a slower speed than that of the bowl. Samples were collected in double in several steps of extraction processes (washed olives, crushed and kneaded pastes, oils from decanter, pomaces), for microbial, chemical and sensory analyses (filtered oils).

2.2. Microbiological Analysis: Enumeration of Yeast Populations

Yeasts were quantified on MYPG agar (malt extract 5 g/L, yeast extract 3 g/L; beef extract 5 g/L, D-glucose 10 g/L) containing sodium propionate (2 g/L) and chloramphenicol (30 mg/mL) in order to inhibit growth of moulds and bacteria, respectively. The samples of olives, pastes and pomaces were plated after decimal dilutions (10 g in 90 mL of physiological saline solution: NaCl, 0.86 g/L homogenized in a Stomacher® 400 (International Pbi, S.P.A., Milano, Italy) for 1 min.

Oil samples from decanter were plated after decimal dilutions (10 mL in 90 mL of physiological saline solution) or by filtration of 10 mL and subsequent washings with physiological solution through 0.45-μm cellulose membranes (Pall Corporation). Yeast colonies were counted after incubation for 48–72 h at 30 °C under aerobic conditions.

2.3. Chemical and Sensory Analyses of Olive Oil

The volatile compounds content was determined according to the literature [30], using solid phase microextraction of the headspace, coupled with a gas chromatograph with a mass spectrometer as a detector (HS-SPME-GC-MS technique). Analysis was performed using the Trace CG instrument combined with a Trace DSQ Thermo Finnigan instrument (Fisher Scientific SAS, Illkirch, France). Quantitative analysis was performed using 4-methyl-2-pentanol as an internal standard. Results were expressed as mg of aromatic compound per Kg of oil.

Acidity (expressed as percentage of oleic acid), peroxide value (meq O_2/Kg) and total phenolic concentration (expressed as mg/Kg of gallic acid) were measured according to EU official method (EC Reg. 1989/2003) [31].

Sensory evaluation of olive oil was performed by a panel test according to the EU official method (EU Reg. 1348/2013) [32]. Samples were analyzed by a panel of professional tasters (8 tasters and a panel leader) recognized by MIPAAF (Ministry of Agricultural Policies, Food and Forestry) since 2002. Intensity of sensory defects and "fruity", "bitter" and "pungent" attributes was assessed and expressed as the median of tasters score on a scale ranging from 0 to 10.

2.4. Molecular Identification of Yeasts

From plates of each sample (washed olives, crushed and kneaded pastes, oils from decanter, pomaces) containing about 300 colonies, 20 colonies were purified and yeast isolates were stored in liquid medium containing 50% (*v/v*) glycerol at −80 °C until further use. Molecular identification of yeast isolates was performed by Randomly Amplified Polymorphic DNA (RAPD) analysis using the primer M13 (5′-GAGGGTGGCGGTTCT-3′) or D1/D2 26S rRNA gene sequencing analysis as reported by Mari et al. [24]. Relative frequencies of isolation used to represent yeast species density according to the isolation source, were calculated as the number of isolates belonging to each species divided by the total number of isolates and expressed in percentage.

2.5. Zymogram Screening for Yeast Enzymatic Activities

72 yeast isolates, belonging to the yeast species most frequently found in oil samples from decanter, were screened for enzymatic activities of potential interest in terms of olive oil quality as reported by Romo-Sánchez et al. [16]. The enzymatic activities screened were cellulase, polygalacturonase, ß-glucosidase, peroxidase, and lipase. The substrates used were, respectively, carboxymethylcellulose (CMC), polygalacturonic acid, cellobiose, H_2O_2, and $CaCl_2$/Tween 80 (all purchased to Sigma Aldrich). Each isolate was grown in YPD (yeast extract 10 g/L; peptone 20g/L, D-glucose 20 g/L) broth at 30 °C for 24 h. To check for lipase activity, cultures were inoculated into 0.1% olive oil integrated with 0.01% Tween 80 broth. Cultures for checking cellulase and ß-glucosidase activity were then grown in a yeast nitrogen base (YNB) broth at 30 °C for 6 h under shaking conditions (100 rpm) for consumption of residual carbon source. Aliquots of 5 µL at 10^6 CFU/mL were spotted on agar plates containing YP (yeast extract 10 g/L; peptone 20 g/L, agar 15 g/L) and 1% of each specific substrate as single carbon source. All plates were incubated at 28 °C for 3 days, except for lipase activity for 7 days. The activity was detected for clear halo for polygalactunorase, according to Fernández González et al., [33], appearance of white precipitation areas for lipase, [34] or growth for cellulase and ß-glucosidase [35]. Peroxidase activity was assessed by oxygen bubble production from H_2O_2.

2.6. Yeast Inoculation into Commercial EVOO

Some isolates belonging to the dominant yeast species (*C. adriatica, N. wickeramii, N. molendini-olei, Y. terventina*) and possessing some enzymatic activities were inoculated into a commercial filtered extra-virgin olive oil (EVOO). Each pure isolate was grown in YPD medium until the early stationary phase and then yeast cells were inoculated in order to have a final concentration of 10^6 cell/mL. The inoculated oil and samples without inoculum as control were placed in sterile glass tubes and bottles, in the dark at a temperature of 15 °C for 180 days until microbial and chemical analysis.

2.7. Statistical Analysis

Microbiological determinations, performed in duplicate, were elaborated according to nonparametric ANOVA followed by Bonferroni Test. Differences were reported at a significance level of $p < 0.05$. Principal Component Analysis (PCA) was used to classify samples. Correlation studies between yeast concentration and the volatile compounds content of oil samples were carried out by calculating both Pearson and Spearman rank correlation coefficients (significance level: $\alpha = 0.05$). All the statistical analyses were performed by Statistica 7.0 software package (Stasoft GmbH, Hamburg, Germany).

3. Results

3.1. Yeast Concentrations Occurring in Different Extractive Olive Oil Processes

The yeast populations present in samples of olives as well as of pastes, oil from decanter and pomaces obtained from the extraction processes carried out in the same crop season at the beginning (HD1), in the middle (HD2) and the end (HD3) of harvesting, were quantified (Figure 1).

The yeast concentration in the olives was not statistically different in the three days of sampling showing an average value of $(5.6 \pm 1.9) \times 10^2$ UFC/g. On the contrary, the yeast concentrations in the kneaded pastes, oil from decanter and pomaces of the first harvesting day (HD1) were statistically lower than the values found in samples from the second and/or the third harvesting day (HD2 and HD3, respectively).

A multidimensional map of the yeast concentrations quantified in pastes, oil from decanter and pomaces of the various extraction processes was obtained by PCA. The sample loading and score plots are reported in Figure 2.

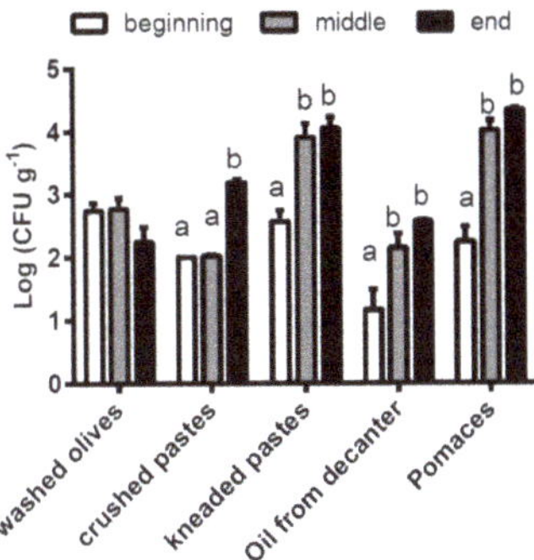

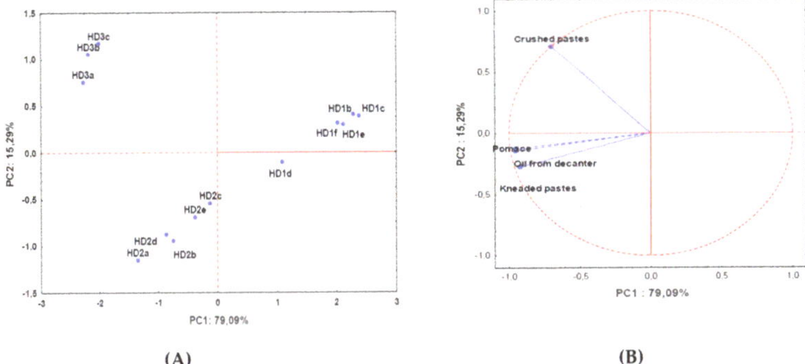

Figure 1. Yeast concentrations at different steps of the oil extraction processes carried out at the beginning (HD1), in the middle (HD2) and the end (HD3) of harvesting in the same crop season. Different letters indicate significant different concentrations within each step (ANOVA, $p < 0.05$).

(A)   **(B)**

Figure 2. Principal Component Analysis of the yeast concentrations in different samples (pastes, oil from decanter and pomaces) during olive oil extraction processes carried out at the beginning (HD1), in the middle (HD2) and the end (HD3) of harvesting in the same crop season. The scores **(A)** and variable loadings **(B)** for the two first principal components.

The model explained 93% of data variability along the first (PC1) and second (PC2) principal components. The extraction processes clustered together according to the same harvesting date. A comparison between the score plot and the loading plot pointed out that the extraction processes showing a higher contamination by yeasts (HD2 and HD3) were all positioned on the left side of the plot. The extraction processes of the third harvesting date (HD3) were located in the left upper quadrant, crushed pastes being characterized by a higher yeast contamination than in the other processes (HD1 and HD2).

3.2. Identification of the Yeast Species

Overall, twelve yeast species belonging to seven genera, besides the yeast-like fungus *Aureobasidium pullulans*, were detected in the different samples collected during the three harvesting days (Table 1). The isolation frequencies of each species were calculated according to the type of sample (olives, pastes, oil, or pomaces) and the harvesting day in which the extraction processes were carried out (Table 1). The comparison among the isolation frequencies highlighted that during the olive oil extractive process some species were typically found in olive fruits whilst other species were associated to crushed and kneaded pastes or found only in oil and in pomaces.

Table 1. Isolation frequencies (%) of yeast species and of the yeast-like fungus *A. pullulans* in different samples collected during olive oil extraction processes carried out at the beginning (HD1), in the middle (HD2), and at the end (HD3) of the same crop season (the symbol "-" indicates isolation frequency < 1%).

Yeast species	Washed Olives			Crushed Pastes			Kneaded Pastes			Oil from Decanter			Pomaces		
	HD1	HD2	HD3	HD1	HD2	HD3	HD1	HD2	HD3	HD1	HD2	HD3	HD1	HD2	HD3
Aureobasidium pullulans	83	100	25	-	-	44	17	-	-	-	-	-	-	-	-
Candida adriatica	-	-	-	-	-	-	-	-	-	-	25	29	-	40	9
Candida diddensiae	-	-	-	-	-	-	6	-	-	-	-	-	-	-	-
Candida kluyveri	-	-	-	-	-	-	53	-	-	-	-	-	-	-	-
Candida norvegica	12	-	50	50	33	13	-	-	-	-	-	-	-	-	-
Nakazawaea wickerhamii	-	-	-	-	-	5	-	4	26	62	24	4	-	9	14
Nakazawaea molendini-olei	-	-	-	-	-	5	6	4	13	13	16	4	-	11	11
Metschnikowia fructicola	-	-	-	-	-	5	-	-	-	25	-	-	-	-	-
Rhodotorula glutinis	-	-	25	-	-	9	-	-	-	-	-	-	-	-	-
Rhodotorula mucilaginosa	-	-	-	50	77	-	-	-	-	-	-	-	11	-	-
Saccharomyces cerevisiae	-	-	-	-	-	-	18	-	8	-	-	-	-	-	-
Yamadazyma terventina	-	-	-	-	-	-	-	-	-	-	35	59	-	5	16
Zygotorulaspora mrakii	-	-	-	-	-	-	-	87	53	-	-	4	55	31	50
Others	5	-	-	-	-	19	-	5	-	-	-	-	34	4	-

Moreover, frequencies of yeast species in different samples varied with the harvesting day. Indeed, washed olives were characterized by a significant presence of three different species: *A. pullulans*, *Candida norvegica*, and *Rhodotorula glutinis* (Table 1) which were often below the detection threshold in other samples. In the crushed pastes *Candida norvegica* and *Rhodotorula mucilaginosa* attained higher percentages than the other species that occurred at frequencies below 1% in samples obtained the first (HD1) and the second (HD2) harvesting day, while in the samples processed on the third harvesting day *A. pullulans* was found at 44%. In the kneaded pastes, the predominant yeast species on the first harvesting day were *Candida kluyveri* and *Saccharomyces cerevisiae* whereas on the second and the third harvesting day *Zygotorulaspora mrakii* and *Nakazawaea wickerhamii*. In oil samples from decanter the yeast species showing significant isolation frequencies were: *Candida adriatica*, *Nakazawaea wickerhamii*, *Nakazawaea molendini-olei*, *Yamadazyma terventina*, and *Metschnikowia fructicola*, although the latter species was isolated only from oil samples of the first harvesting day and *Y. terventina* and *C. adriatica* only from oil samples of the second and third harvesting day. Finally, pomaces were characterized by the presence of the same yeast species isolated from oil samples with the exception of *Z. mrakii*, a species isolated mainly from kneaded pastes of the second and third harvesting day.

3.3. Chemical and Sensory Characteristics of Olive Oil Samples

Chemical and sensory analyses of olive oil samples obtained from different extractive processes were performed for oil quality assessment (Table S1 and Table S2). The concentrations of 48 volatile compounds in the 14 olive oil samples were quantified and used to obtain a multidimensional map by PCA, with the exception of data related to oil obtained from the first extractive process (HD1a). Indeed, the plant usually works about three months a year and, therefore, data obtained from the first extractive process might be affected by the environmental conditions occurred during the stopping time and, thus, not be representative. The relevant sample loading and score plots are reported in Figure 3.

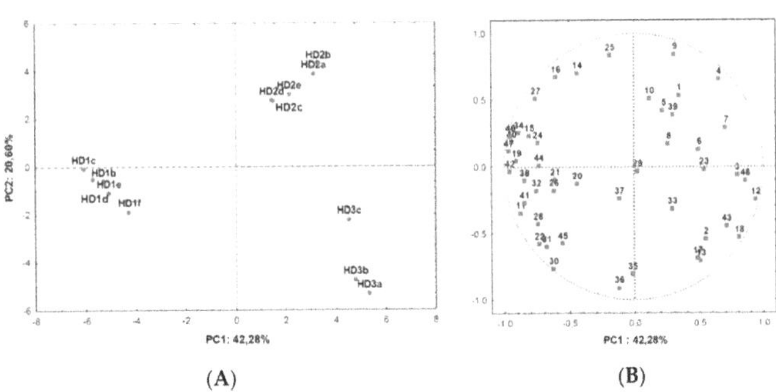

Figure 3. Principal Component Analysis carried out on volatile compounds content of olive oil samples produced during different extractive processes (a, b, c, d, f) at the beginning (HD1), in the middle (HD2) and at the end (HD3) of the same crop season. The scores (**A**) and variable loadings (**B**) for the two first principal components. Variables: (1) Heptane; (2) Octane; (3) Methyl acetate; (4) Ethyl acetate; (5) 2-Butanone; (6) 2-Methyl-butanal; (7) Isovaleraldehydes; (8) Valeraldehydes; (9) Ethyl-vinyl-ketone; (10) Propanol; (11) Hexanal; (12) Isobutanol; (13) 2-Pentanol; (14) trans-2-Pentenal; (15) cis-3-Hexenal; (16) 1-Penten-3-ol; (17) 2-Heptanone; (18) 2 and 3-Methylbutan-1-ol; (19) trans-2-Hexenal; (20) Ocimene; (21) Pentanol; (22) Hexyl acetate; (23) 2-Octanone; (24) Octanal; (25) trans-2-Pentenol; (26) cis-3-Hexenyl acetate; (27) cis-2-Pentenol; (28) trans-2- Hexenyl acetate; (29) 6-Methyl-5-epten-2-one; (30) Hexanol; (31) trans-3-Hexen-1-ol; (32) cis-3-Hexenol; (33) Nonanal; (34) 2,4-Exadienal; (35) trans-2-Hexenol; (36) cis-2-Exenol; (37) trans-2-Octanal; (38) 1-Octen-3-ol; (39) 2,4-Heptadienal; (40) Benzaldehyde; (41) Octanol; (42) Butyric acid; (43) trans-2-Decenal; (44) Nonanol; (45) Ethylbenzene (46) Phenol; (47) 4-Ethylphenol; (48) 1-Penten-3-one.

The model explained 63% of data variability along the first (PC1) and second (PC2) principal components. All the assayed oil samples clustered according to the harvesting date of the olives. The oils of the first harvesting date were significant different respect to the other oils, being characterized by high values of: 1-penten-3-ol, cis-3-hexenal, cis-3-hexenyl acetate, cis-2-penten-1-ol, trans-2-hexenyl acetate. On the contrary, the most olive oil samples extracted from olives of the second harvesting date contained high concentrations of ethyl vinyl ketone, 2-butanone, propanol, heptane and 2,4-heptadienal. Finally, the oil samples of third harvesting date were characterized by high values of methyl acetate, isobutanol, 2 and 3-methylbutan-1-ol, trans-2-decenal, octane, 2-heptanone, 2-pentanol. The olive oil samples obtained from olives of the first harvesting date (HD1b, HD1c, HD1d, HD1e, and HD1f) and by processes with the lowest level of yeast contamination (Figure 2), grouped together on the left side of the plot (Figure 3). In contrast, the oil samples produced in the middle and at the end of harvesting were positioned on the right side of the plot.

In summary, as the olive harvest proceeded, the oil flavour changed, going e.g. from grassy to more buttery notes as it was shown in Figure 3 considering H1 and H3 samples.

In order to investigate on the possible relation between yeast concentrations found in kneaded pastes or in oil from decanter and the concentrations of volatile compounds in olive oils, correlation studies were carried out and the results are reported in Table 2.

Table 2. Statistically significant correlations ($p < 0.05$) calculated between yeast concentrations occurring in kneaded pastes or oil from decanter and volatile compounds of the final olive oil samples. (ns = not significant).

Compounds	Kneaded Pastes		Oil from Decanter	
	Spearman r	Pearson r	Spearman r	Pearson r
Aldehydes				
Hexanal	−0.6998	ns	−0.7660	−0.6321
cis-3-Hexenal	−0.6203	−0.633	−0.7660	−0.8160
trans-2-Hexenal	−0.5982	−0.603	−0.6115	−0.5791
2-Methyl-butanal	0.6556	ns	0.6733	0.6797
Isovaleraldehydes	0.9161	0.7170	0.8808	0.8371
2,4-Heptadienal	−0.7086	−0.6413	−0.7704	−0.7863
Benzaldehydes	−0.7572	ns	−0.7616	−0.5704
trans-2-Decenal	0.5378	0.5590	ns	ns
Esters				
Methyl acetate	0.8013	0.7186	0.8013	0.7650
Ethyl acetate	0.7572	0.6547	0.7130	0.5935
Hexyl acetate	−0.7925	−0.6573	−0.8499	−0.8145
trans-2- Hexenyl acetate	−0.7339	−0.8582	−0.7251	−0.7072
cis-3-Hexenyl acetate	−0.7484	−0.6344	−0.6954	−0.8006
4-Ethyl-phenol	−0.7042	−0.5673	−0.7439	−0.6407
Carboxylic acids and ketones				
Butyric acid	−0.6733	−0.6317	−0,691	−0.6885
1-Penten-3-one	0.8102	0.7454	0.713	0.7172
Alcohols				
Isobutanol	0.8318	0.7596	0.7628	0.7562
2 and 3-Methylbutan-1-ol	0.8013	0.6240	0.7704	0.6024
trans-3-Hexen-1-ol	−0.7307	−0.6868	−0.7042	−0.7060
cis-3-Hexenol	−0.7660	−0.6455	−0.7881	−0.8419
Nonanol	−0.5938	−0.4505	−0.5717	ns
Hexanol	−0.7484	−0.6560	−0.6954	−0.6181
1-Octen-3-ol	−0.6556	ns	−0.6998	−0.5398

3.4. Enzymatic Activity of Yeasts

Yeasts belonging to the species most frequently isolated from decanter oil samples (*C. adriatica*, *N. wickeramii*, *N. molendini-olei*, *Y. terventina*) and coming from different extraction processes were assayed for their enzymatic capabilities with the aim to verify if they could potentially influence the chemical composition of the olive oil. The results are shown in Table 3. All the isolates displayed peroxidase activity, on the contrary no isolates showed cellulase or polygalacturonase activity. All the isolates belonging to *N. molendini-olei* species showed high ß-glucosidase activity, while in the other species this enzymatic activity resulted strain-dependent. Lipase activity was absent in all the isolates of *N. molendini-olei* and present in only one isolate of *N. wickerhamii* species. On the contrary, all isolates of *C. adriatica* and *Y. terventina* species displayed lipase activity. In particular, almost 50% of *Y. terventina* isolates showed high levels of this enzymatic activity.

Table 3. Enzymatic activities of the yeast species isolated from olive oil samples obtained from decanter in different extraction processes.

Yeast Species	N. of Isolates	Enzymatic Activity									
		ß-glucosidase				Lipase				Peroxidase	
		-	+	++	+++	-	+	++	+++	-	+
C. adriatica	13	0	1	3	9	0	4	8	1	0	13
N. molendini-olei	23	0	0	0	23	23	0	0	0	0	23
N. wickerhamii	21	4	0	4	13	20	0	0	1	0	21
Y. terventina	15	6	5	3	1	0	6	1	8	0	15

3.5. Yeast Inoculation and Chemical Composition of Extra Virgin Olive Oil (EVOO)

Three strains, belonging to the species most frequently isolated from the extractive processes and detected also in extra virgin olive oil during conservation (*N. molendini-olei* PG194, *N. wickerhamii* DM15, and *Y. terventina* DFX3), were chosen to test whether their enzymatic activities were displayed in EVOO. The three yeast strains showed peroxidase activity, high ß-glucosidase activity and only two of them also high lipase activity (*N. wickerhamii* DM15 and *Y. terventina* DFX3). Isolates of *C. adriatica* were not considered because this species was not frequently found in Tuscan olive oil during conservation. The oil used in the trials was a filtered five-month-old extra virgin olive oil showing a yeast concentration below 10 CFU/mL. To maximize the yeast effect of each species, the three yeast strains were separately inoculated in oil to obtain a final concentration of 10^6 CFU/mL. During storage, the yeast cells viability in the olive oil decreased according to the isolate inoculated. In detail, the suspended living cells recovered from the samples after two months of storage varied from a minimum of 10^2 CFU/mL, observed in the olive oil inoculated with *N. molendini-olei*, to a maximum of 10^3 CFU/mL found in the oil samples inoculated with *N. wickerhamii* and *Y. terventina*. The analytical indices of treated olive oil evaluated after two months of storage, showed some statistical different results for free fatty acids (% oleic acid), peroxide value and total polyphenols (ANOVA, $p < 0.05$). More specifically, the free fatty acids of inoculated oils with *N. wickerhamii* and *Y. terventina* reached values (both 0.28 ± 0.02 % of oleic acid) significantly higher than the control (un-inoculated oil) and the oil inoculated with *N. molendini-olei* (both 0.25 ± 0.01 % of oleic acid). The peroxide values were higher than the control (13.74 ± 0.38 meq O_2/Kg) only in the olive oil inoculated with *Y. terventina* (16.45 ± 1.41 meq O_2/Kg). Finally, total polyphenols were 10% lower than that in the control (650 ± 35 mg/kg). In any case, all the inoculated olive oils retained the requirements of extra virgin oil.

Volatile compounds content of the control and the inoculated oils were quantified and used to obtain a multidimensional map by PCA. The relevant sample loading and score plots are reported in Figure 4. The model explained 83% of data variability along the first (PC1) and second (PC2) principal components.

A comparison between the score plot and the loading plot showed that the control was significant different respect to the inoculated oils, which were all positioned on the right side of the plot. Significant differences were also observed between oils inoculated with different yeasts isolates, in particular between *Y. terventina* DFX3 and *N. molendini-olei* PG194.

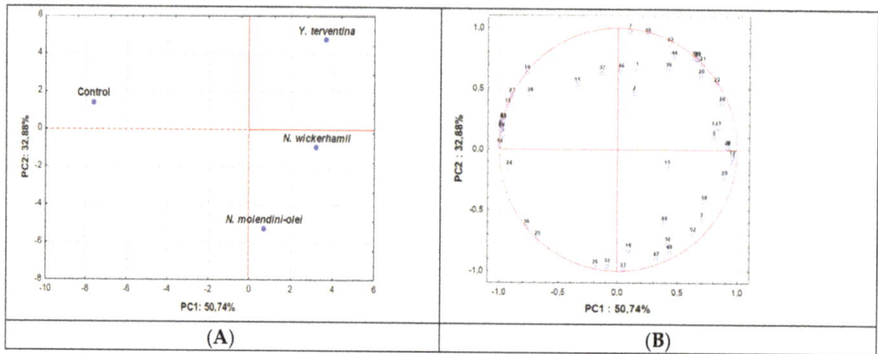

Figure 4. Principal Component Analysis carried out on volatile compounds content of oils inoculated or not (control) with different yeast isolates. The scores (A) and variable loadings (B) for the two first principal components. Variables: (1) 1-Octen-3-one; (2) 1-Penten-3-ol; (3) 2,4-Decadienal; (4) 2,4-Heptadienal; (5) trans, trans-2,4-Nonadienal; (6) Butan-2-one; (7) 2 and 3-Methylbutan-1-ol; (8) 2-Methylbutanal; (9) Octan-2-one; (10) 1-Penten-3-one; (11) 4-Ethylphenol; (12) 6-Methyl-5-hepten-2-one; (13) trans-2-Decenal; (14) trans-2-Heptenal; (15) trans-2-Hexenal; (16) trans-2-Hexenyl acetate; (17) trans-2-Hexen-1-ol; (18) trans-2-Pentenal; (19) trans-2-Pentenol; (20) trans-2-Hexen-1-ol; (21) Phenol; (22) cis-2-Pentenol; (23) cis-3-Hexenal; (24) cis-3-Hexenyl acetate; (25) cis-3-Hexenol; (26) Butyric acid; (27) Heptanoic acid; (28) Octanoic acid; (29) Pentanoic acid; (30) Propanoic acid; (31) Heptanal; (32) Heptane; (33) Heptan-2-ol; (34) Hexanal; (35) Hexanol; (36) Hexyl acetate; (37) Ethyl acetate; (38) Ethyl isobutyrate; (39) Ethylguaiacol; (40) Ethyl propionate; (41) Ethyl vinyl ketone; (42) Phenylethanol; (43) Guaiacol; (44) Isobutanol; (45) Isovaleraldehydes; (46) Methyl acetate; (47) Ethyl propionate; (48) Nonanol; (49) Octanal; (50) Octane; (51) Octan-2-ol; (52) Pentanol; (53) Propanol; (54) Valeraldehydes.

4. Discussion

The yeast concentrations occurring in olive oil extraction processes significantly increased from the first to the last harvesting day of the same crop season. Moreover, the occurrence of different yeast species according to the date of sampling (beginning, middle, and end) demonstrated the progressive contamination of the extraction plant that selects some yeast species at the expense of others. This is the case of some yeast species (as *Z. mrakii* in kneaded pastes as well as *Y. terventina* and *C. adriatica* in oil from decanter) that, being below the detection threshold in the first harvesting date, were then detected at significant level in the second and third harvesting date. The yeast species isolated were in agreement with the results obtained from other surveys carried out on oleic ecosystem [13,16,17,21,27]. Despite all the oil samples were classified as extra virgin olive oils (EU Reg. 1348/2013) [32]., the level of yeast-contamination of the various processes seemed to affect the olive oil chemical composition. Other Authors [17,20,22,23,36] found that some yeast species affect the organoleptic properties of virgin olive oil but no relationships with yeast population concentrations were detected. The oils obtained from processes characterized by a lower yeast contamination (first harvesting date), were characterized by higher concentrations of compounds mostly related to olive oil positive attribute such as "fruity" [1,37,38]. Among these compounds are included cis-3-hexenal and cis-3-hexenyl acetate that are associated with sensory descriptor "Green" [37,38], while cis-2-penten-1-ol to "banana" [37].

On the contrary, the oils obtained from processes more contaminated by yeasts (especially from the third harvesting date) contained higher concentrations of molecules, which were often related to negative attribute [1,3]. Among these compounds were included 2 and 3-methylbutan-1-ol and methyl acetate, 2 and 3-methylbutan-1-ol that are associated with sensory descriptor "winey" and "woody" respectively and both involved in "Mustiness-humidity", "Fusty" and "Winey-vinegary" negative attribute [3]. Methyl acetate is related to "Winey-vinegary or acid-sour" defect, while trans-2-decenal (sensory descriptor: "painty", "fishy", "fatty") to "Rancid" [1,37,38].

Finally, correlation studies between yeast concentrations in kneaded pastes or oil from decanter and the volatile compounds of the final oils demonstrated significant positive correlations with compounds (i.e. methyl acetate, ethyl acetate, 2 and 3-methylbutan-1-ol) related to "Winey-vinegary or acid-sour" defects. At the same time, significant negative correlations with compounds related to positive attribute were observed (i.e. hexyl acetate; cis-3-hexenyl acetate; trans-2-hexenyl acetate; cis-3-hexenol; 2,4-hexadienal) [1]. In other words, the greater is the yeasts contamination occurring in olive oil extraction processes and the worse is the organoleptic quality of the oil. The only exception was represented by butyric acid, usually related to "rancid" defects [1,3] that was negatively correlated with the yeasts concentration both in kneaded pastes and oil from decanter.

Olive oil chemical characteristics may be affected in different way depending on the enzymatic capabilities of the yeast microbiota occurring in olive oil extraction processes. In fact, most of the enzymatic activities able to modify the olive oil chemical composition were species or strain-dependent as generally reported [1,15,16,24,39].

In this study, peroxidase activity, responsible of a negative influence on olive oil quality due to oxidative degradation of the protective phenol compounds [40] was common to all the species assayed. On the contrary, cellulase and polygalacturonase activities that increase antioxidant phenol compound levels conferring a protective effect by hydrolysing olive cell-wall polysaccharides [41], were absent in all the assayed isolates. Finally, β-glucosidase and lipase activities were strain and/or species-dependent. The β-glucosidase enzyme is involved in the degradation of oleuropeine into a heterosidic ester of elenolic acid and 3,4-dihydroxyphenylethanol; both of these compounds are technologically important in view of their browning capacity and intense bitter taste [11,12,42].

Lipase activity can impair product quality due to the increase of both the diglyceride and acidity levels through hydrolysis of triacylglycerols [24,43]. Considering that the olives are fruits with high fat concentrations, the presence of lipolytic yeasts in olive oil could modify the nutritional composition and organoleptic characteristic of this product.

When three representative strains (*N. molendini-olei* PG194, *N. wickerhamii* DM15, *Y. terventina* DFX3) characterized by different enzymatic capabilities were inoculated in olive oil, different effects on oil chemical composition were detected. The analytical indices, used to classify an olive oil as extra-virgin, showed significant differences: the acidity level increased when *C. wickeramii* DM15 and *Y. terventina* DFX3 were present; peroxide values increased only in the presence of *Y. terventina* DFX3, total polyphenols decreased independently of the inoculated yeast strain.

Finally, also the volatile compounds content that resulted were strongly influenced by the yeast strain inoculated.

To generalize, in the samples of oil treated with yeasts, a higher concentration of some compounds responsible of negative oil attributes (i.e.: trans 2-heptenal, 6-methyl-5-hepten-2-one, 2-octanone) and a lower concentration of C6 volatile carbonyl compounds responsible for positive oil attributes, were found. Similarly, Zullo et al. [21] observed a lower content of C6 volatile carbonyl compounds when a *N. wickerhamii* strain was inoculated in oil.

5. Conclusions

In conclusion, during the olive oil crop season, some yeast species colonize the extraction plant (malaxation equipment and decanter in particular) at the expense of others becoming the dominant microbiota. This colonization significantly affects the volatile compound content of the olive oils; indeed, the oils obtained in the first days of the olive oil crop season were significantly different from the others. The effects of the yeasts colonization on the chemical characteristics of the oils depend on not only by the population density but also by the enzymatic capabilities of the species and/or the strains composing the microbiota. Therefore, the hygienic condition of the olive oil extraction plant is important in the definition of an olive oil aromatic profile. In this contest, it could be of interest to investigate if each olive oil extraction plant might select a typical microbiota, with metabolic capabilities potentially able to affect in a characteristic way the aromatic composition of the final product.

Supplementary Materials: The following are available online at http://www.mdpi.com/2304-8158/8/10/457/s1, Table S1: Chemical analyses of olive oil samples obtained from different extractive processes (a, b, c, d, f) carried out at the beginning (HD1), in the middle (HD2) and the end (HD3) of harvesting in the same crop season (U = measurement uncertainty), Table S2: Volatile compounds (mg/kg) in olive oil samples obtained from different extractive processes (a, b, c, d, f) carried out at the beginning (HD1), in the middle (HD2) and the end (HD3) of harvesting in the same crop season.

Author Contributions: Conceptualization: S.G., E.M., D.B.; Investigation: D.B., E.M.; Resources: D.B.; Data curation, S.G., D.B., E.M.; Writing—Original draft preparation, L.G., S.G., E.M; Writing, review and editing, L.G., S.G.; Supervision, L.G, S.G.

Funding: This research received no external funding.

Conflicts of Interest: The authors declare no conflict of interest.

References

1. Zanoni, B. Which processing markers are recommended for measuring and monitoring the transformation pathways of main components of olive oil? *Ital. J. Food Sci.* **2014**, *26*, 3–11.

2. Aparicio, R.; Morales, M.T. Characterization of olive ripeness by green aroma compounds of virgin olive oil. *J. Agric. Food Chem.* **1998**, *46*, 1116–1122. [CrossRef]

3. Morales, M.T.; Luna, G.; Aparicio, R. Comparative study of virgin olive oil sensory defects. *Food Chem.* **2005**, *91*, 293–301. [CrossRef]

4. Bendini, A.; Valli, E.; Barbieri, S.; Gallina Toschi, T. Sensory analysis of virgin olive oil. Ch. 6; In *Olive Oil—Constituents, Quality, Health Properties and Bioconversions*; Boskou, D., Ed.; InTech Open Access Publisher: London, UK, 2012; p. 109. ISBN 978-953-307-921-9.

5. Campestre, C.; Angelini, G.; Gasbarri, C.; Angerosa, F. The compounds responsible for the sensory profile in monovarietal virgin olive oils. *Molecules* **2017**, *22*, 1833. [CrossRef] [PubMed]

6. Taticchi, A.; Esposto, S.; Servili, M. The basis of the sensory properties of virgin olive oil.olive oil sensory science. Ch. 2; In *Olive Oil Sensory Science*, 1st ed.; Monteleone, E., Langstaff, S., Eds.; John Wiley & Sons Ltd: Oxford, UK, 2014.

7. Vichi, S.; Romero, A.; Tous, J.; Caixach, J. The activity of healthy olive microbiota during virgin olive oil extraction influences oil chemical composition. *J. Agr. Food Chem.* **2011**, *59*, 4705–4714. [CrossRef]

8. Angerosa, F.; Campestre, C. Sensory quality: methodologies and applications. In *Handbook of Olive Oil-Analysis and Properties*; Aparicio, R., Harwood, J., Eds.; Springer: New York, NY, USA, 2013; pp. 523–560.

9. Aparicio, R.; Morales, M.T.; Aparicio-Ruiz, R.; Tena, N.; García-González, D.L. Authenticity of olive oil: Mapping and comparing official methods and promising alternatives. *Food Res. Int.* **2013**, *54*, 2025–2038. [CrossRef]

10. Guerrini, S.; Mari, E.; Migliorini, M.; Cherubini, C.; Trapani, S.; Zanoni, B.; Vincenzini, M. Investigation on microbiology of olive oil extraction process. *Ital. J. Food Sci.* **2015**, *27*, 237–247. [CrossRef]

11. Ciafardini, G.; Zullo, B.A. Microbiological activity in stored olive oil. *Int. J. Food Microbiol.* **2002**, *75*, 111–118. [CrossRef]

12. Ciafardini, G.; Zullo, B.A. Survival of microorganisms in extra virgin olive oil during storage. *Food Microbiol.* **2002**, *19*, 105–109. [CrossRef]

13. Zullo, B.A.; Cioccia, G.; Ciafardini, G. Distribution of dimorphic yeast species in commercial extra virgin olive oil. *Food Microbiol.* **2010**, *27*, 1035–1042. [CrossRef]

14. Ciafardini, G.; Zullo, B.A.; Cioccia, G.; Iride, A. Lipolytic activity of *Williopsis californica* and *Saccharomyces cerevisiae* in extra virgin olive oil. *Int. J. Food Microbiol.* **2006**, *107*, 27–32. [CrossRef] [PubMed]

15. Ciafardini, G.; Zullo, B.A.; Iride, A. Lipase production by yeasts from extra virgin olive oil. *Food Microbiol.* **2006**, *23*, 60–67. [CrossRef] [PubMed]

16. Romo-Sánchez, S.; Alves-Baffi, M.; Arévalo-Villenaa, M.; Úbeda-Iranzoa, J.; Briones-Pérez, A. Yeast biodiversity from oleic ecosystems: Study of their biotechnological properties. *Food Microbiol.* **2010**, *27*, 487–492. [CrossRef] [PubMed]

17. Ciafardini, G.; Zullo, B.A. Effect of lipolytic activity of *Candida adriatica*, *Candida diddensiae* and *Yamadazyma terventina* on the acidity of extra-virgin olive oil with a different polyphenol and water content. *Food Microbiol.* **2015**, *47*, 15–20. [CrossRef] [PubMed]

18. Palomares, O.; Villalba, M.M.; Rodriguez, R. The C-terminal segment of the 1,3-β-glucanase Ole e 9 from olive (*Olea europea*) pollen is an independent domain with allergenic activity: Expression in *Pichia pastoris* and characterization. *Biochem. J.* **2003**, *369*, 593–601. [CrossRef] [PubMed]

19. Migliorini, M.; Cecchi, L.; Cherubini, C.; Trapani, S.; Cini, E.; Zanoni, B. Understanding degradation of phenolic compounds during olive oil processing by inhibitor addition. *Eur. J. Lipid Sci. Technol.* **2012**, *114*, 942–950. [CrossRef]

20. Palla, M.; Digiacomo, M.; Cristani, C.; Bertini, S.; Giovannetti, M.; Macchia, M.; Manera, C.; Agnolucci, M. Composition of health-promoting phenolic compounds in two extra virgin olive oils and diversity of associated yeasts. *J. Food Comp. Anal.* **2018**, *74*, 27–33. [CrossRef]

21. Zullo, B.A.; Cioccia, G.; Ciafardini, G. Effects of some oil-born yeasts on the sensory characteristics of Italian virgin olive oil during its storage. *Food Microbiol.* **2013**, *36*, 70–78. [CrossRef]

22. Zullo, B.A.; Ciafardini, G. Changes in physicochemical and microbiological parameters of short and long-lived veiled (cloudy) virgin olive oil upon storage in the dark. *Eur. J. Lipid Sci. Technol.* **2018**, *120*, 1–8. [CrossRef]

23. Zullo, B.A.; Ciafardini, G. Virgin olive oil yeasts: A review. *Food Microbiol.* **2018**, *70*, 245–253. [CrossRef]

24. Zullo, B.A.; Ciafardini, G. Lipolytic yeast distribution in commercial extra virgin olive oil. *Food Microbiol.* **2008**, *25*, 970–977. [CrossRef] [PubMed]

25. Čadež, N.; Raspor, P.; Turchetti, B.; Cardinali, G.; Ciafardini, G.; Veneziani, G.; Peter, G. Candida adriatica sp. nov. and Candida molendinolei sp. nov., two novel yeast species isolated from olive oil and its by-products. *Int. J. Syst. Evol. Microbiol.* **2012**, *62*, 2296–2302. [CrossRef] [PubMed]

26. Čadež, N.; Dlauchy, D.; Raspor, P.; Péter, G. *Ogataea kolombanensis* sp. nov., *Ogataea histrianica* sp. nov. and *Ogataea deakii* sp. nov., three novel yeast species from plant sources. *Int. J. Syst. Evol. Microbiol.* **2013**, *63*, 3115–3123. [CrossRef] [PubMed]

27. Ciafardini, G.; Zullo, B.A.; Antonielli, L.; Corte, L.; Roscini, L.; Cardinali, G. *Yamadazyma terventina* sp.nov., a yeast species of the *Yamadazyma* clade from Italian olive oils. *Int. J Syst. Evol. Microbiol.* **2013**, *63*, 372–376. [CrossRef] [PubMed]

28. Mari, E.; Guerrini, S.; Granchi, L.; Vincenzini, M. Yeast microbiota in the olive oil extractive process: A three-year study at an industrial scale. *World J. Microbiol. Biotechnol.* **2016**, *32*, 93–103. [CrossRef] [PubMed]

29. Ciafardini, G.; Cioccia, G.; Zullo, B.A. Taggiasca extra virgin olive oil colonization by yeasts during the extraction process. *Food Microbiol.* **2017**, *62*, 58–61. [CrossRef] [PubMed]

30. Vichi, S.; Pizzale, L.; Conte, L.S.; Buxaderas, S.; Lopez-Tamames, E. Solid-phase microextraction in the analysis of virgin olive oil volatile fraction: Characterization of virgin olive oils from two distinct geographical areas of northern Italy. *J. Agric. Food Chem.* **2003**, *51*, 6572–6577. [CrossRef] [PubMed]

31. *Commission Regulation (EC) No 1989/2003. OJEU. L295:57*; Publications Office of the European Union: Luxembourg, 2003.

32. *Commission Implementing Regulation (EU) No 1348/2013. OJEU. L338:31*; Publications Office of the European Union: Luxembourg, 2013.

33. Fernández González, M.; Úbeda, J.; Vasudevan, T.G.; Cordero Otero, R.; Briones, A. Evaluation of polygalacturonase activity in Saccharomyces wine strains. *FEMS Microbiol. Lett.* **2004**, *237*, 261–266. [CrossRef] [PubMed]

34. Nuero, O.M.; Garcia-Lepe, R.; Lahoz, M.C.; Santamaria, F.; Reyes, F. Detection of lipase activity on ultrathin-layer isoelectric focusing gels. *Anal. Biochem.* **1994**, *222*, 503–505. [CrossRef]

35. Arévalo Villena, M.; Úbeda, J.F.; Cordero Otero, R.R.; Briones, A.I. Optimisation of a rapid method for studying the cellular location of b-glucosidase activity in wine yeasts. *J. Appl. Microbiol.* **2005**, *99*, 558–566. [CrossRef]

36. Cayuela, J.A.; Gómez-Coca, R.B.; Moreda, W.; Pérez-Camino, M.C. Sensory defects of virgin olive oil from a microbiological perspective. *Trends Food. Sci. Technol.* **2015**, *43*, 227–235. [CrossRef]

37. Angerosa, F.; Servili, M.; Selvaggini, R.; Taticchi, A.; Esposto, S.; Monteodoro, G. Volatile compounds in virgin olive oil: Occurrence and their relationship with the quality. *J. Chromatogr.* **2004**, *A1054*, 17–31. [CrossRef]

38. Kalua, C.M.; Allen, M.S.; Bedgood Jr., D.R.; Bishop, A.G.; Prenzler, P.D.; Robards, K. Olive oil volatile compounds, flavour development and quality: A critical review. *Food Chem.* **2007**, *100*, 273–286. [CrossRef]

39. Bautista-Gallego, J.; Rodríguez-Gómez, F.; Barrio, E.; Querol, A.; Garrido Fernández, A.; Arroyo-López, F.N. Exploring the yeast biodiversity of green table olive industrial fermentations for technological applications. *Int. J. Food Microbiol.* **2011**, *147*, 89–96. [CrossRef] [PubMed]

40. Gomez-Rico, A.; Fregapane, G.; Salvador, M.D. Effect of cultivar and ripening on minor components in Spanish olive fruits and their corresponding virgin olive oils. *Food Res. Int.* **2008**, *41*, 433–440. [CrossRef]

41. De Faveri, D.; Aliakbarian, B.; Avogadro, M.; Perego, P.; Converti, A. Improvement of olive oil phenolics content by means of enzyme formulations: Effect of different enzyme activities and levels. *Biochem. Eng. J.* **2008**, *41*, 149–156. [CrossRef]

42. Ciafardini, G.; Marsilio, V.; Lanza, B.; Pozzi, N. Hydrolysis of oleuropein by *Lactobacillus plantarum* strains associated with olive fermentation. *Appl. Environ. Microbiol.* **1994**, *60*, 4142–4147.

43. Cardenas, F.; Alvarez, E.; de Castro-Alvarez, M.S.; Sánchez-Montero, J.M.; Valmaseda, M.; Élson, S.W.; Sinisterra, J.V. Screening and catalytic activity in organic synthesis of novel fungal and yeast lipases. *J. Mol. Catal. B Enzym.* **2001**, *14*, 111–123. [CrossRef]

MDPI
St. Alban-Anlage 66
4052 Basel
Switzerland
Tel. +41 61 683 77 34
Fax +41 61 302 89 18
www.mdpi.com

Foods Editorial Office
E-mail: foods@mdpi.com
www.mdpi.com/journal/foods